System-on-Chip Architectures and Implementations for Private-Key Data Encryption

Máire McLoone and **John V. McCanny**
Queen's University
Belfast, Northern Ireland

Kluwer Academic/Plenum Publishers
New York, Boston, Dordrecht, London, Moscow

Library of Congress Cataloging-in-Publication Data

McLoone, Máire.
  System-on-chip architectures and implementations for private-key data encryption/
  Máire McLoone, John V. McCanny.
      p.   cm.
  Includes bibliographical references and index.
  ISBN 0-306-47882-X
      1. Computer security.   2. Data encryption (Computer science).   3. Cryptography.   4.
  Computer architecture.   I. McCanny, J. V.   II. Title.

QA76.9.A25M435 2003
005.8—dc22
                                                                                    2003054468

ISBN 0-306-47882-X

©2003 Kluwer Academic / Plenum Publishers, New York
233 Spring Street, New York, New York 10013

http://www.wkap.com

10 9 8 7 6 5 4 3 2 1

A C.I.P. record for this book is available from the Library of Congress

All rights reserved

No part of this book may be reproduced, stored in a retrieval system, or transmitted in any form
or by any means, electronic, mechanical, photocopying, microfilming, recording, or otherwise,
without written permission from the Publisher, with the exception of any material supplied
specifically for the purpose of being entered and executed on a computer system, for exclusive
use by the purchaser of the work

Permissions for books published in Europe: *permissions@wkap.nl*
Permissions for books published in the United States of America: *permissions@wkap.com*

Printed in the United States of America

Acknowledgments

I would like to express my thanks to Amphion Semiconductor Ltd., the European Social Fund and Queen's University Belfast for the financial support provided to me during my PhD research work, on which this book is based.

Thanks to all my colleagues in the DSP and Telecommunications group for their help and friendship. Thanks also to Richard Ruddock for providing excellent technical support and to Paula Dougherty and Laurence Downey for all their help.

A very special thanks to Shane for his help, patience, love and friendship.

Finally, I am indebted to my Mum and Dad, my brothers, Seán and Séamus, and sisters, Eibhlín and Nábla, for their love and support over the years.

Máire McLoone

Foreword

It is a pleasure and an honor to write a foreword for this book.

In the (quite recent) past, personal and sensitive data and the computations on these data were centralized within a well-described and physically protected environment.

Since then, the world has evolved to one where data and computations are distributed over a wide range of computing devices, connected together through wired and wireless links. This has ignited a revolution in security issues and protocols: how to protect privacy, authentication, integrity, and so on in this distributed connected computing world.

Key to this are efficient implementations of the cryptographic algorithms that support these security protocols. Here lies the contribution of this book: it is one of the first ones, if not the first, which focuses on efficient hardware implementation strategies for secret key algorithms. AES and DES, the main secret-key algorithms in use today are the core contributions of this work, not from a mathematical viewpoint, but from an implementation viewpoint. Secondly, the focus is on efficient architectures and design strategies for FPGA and ASIC realizations. The optimizations for these platforms are very different from traditional software implementations. This makes this book of a very practical use to any engineer in need for efficient realizations of AES and DES.

Ingrid Verbauwhede, Ph.D.

Preface

In recent years there has been an increased awareness of Information Technology (IT) security-related issues. Personal computers are no longer used exclusively in the office – the use of home and recreational computers has increased dramatically and the majority of these have Internet access to resources such as email, newsgroups and on-line shopping and banking. However, this widespread availability and acceptance of computers has also increased the number of people with an ability to compromise data. This has led to a very high percentage of traffic that requires safeguarding. One of the most fundamental and widespread tools used in providing Internet security is encryption. In an effort to increase consumer confidence in electronic transactions, encryption techniques are now in widespread use and are in high demand. Cryptography has become one of the main tools for providing privacy, trust, access control, secure electronic payments and in general, secure data communication. This need for high-strength encryption will continue as there is an overwhelming demand for new and improved communication technologies. For example, a fast growing area of interest is that of Wireless Local Area Networks (WLANs) for which security is an essential aspect. Due to their wireless nature WLANs are more vulnerable than their wired counterparts and as a result of this inherent vulnerability data security is crucial.

The rapid developments in communication systems have resulted in a need to perform encryption on data in real-time. Encryption of digital information in real-time holds the key to the successful growth of major applications in areas such as satellite communications and electronic commerce. Cryptography is typically a mathematically intensive process which, to date, has mainly been implemented in software. However, such methods are slow and cannot cope with the demands of rapidly growing broadband communication systems. Therefore, innovative hardware solutions involving mapping of complex mathematical operations onto special purpose silicon circuit architectures provide the only feasible solution.

Secure communications systems often require the capacity to encrypt messages with several different algorithms in addition to the need to change keys regularly. The concept of re-usable, parameterisable cryptographic designs is ideally suited to meeting these requirements. Re-usable silicon designs of this nature are known as Intellectual Property (IP) cores. IP cores are also an integral part of System-on-Chip (SoC) design. This emerging technique involves the integration of pre-designed IP cores onto a single chip in order to implement highly complex applications.

In this book, new generic silicon architectures for the Data Encryption Standard (DES) and the Rijndael symmetric key encryption algorithms are presented. DES, the US Federal Information Processing Standard (FIPS) for over twenty years, was replaced by the Rijndael algorithm in November 2001. The generic architectures can be utilised to rapidly and effortlessly generate cores, which support numerous application requirements, most importantly, different modes of operation and encryption and decryption capabilities. Hash algorithms, which provide authentication, are an essential element of any cryptographic application. Authentication offers assurance about the content and origin of communicated messages. Efficient silicon architectures of the SHA-1, SHA-384, SHA-512 and HMAC-SHA-1 hash functions are also described. The Internet Protocol Security (IPSec) standard is an application which employs both encryption and authentication. A single-chip IPSec architecture is presented that comprises the generic Rijndael design and a highly efficient HMAC-SHA-1 implementation. This architecture can also be utilised to provide the encryption and authentication needs of applications such as Wireless Local Area Networks (WLANs) and the Secure Electronic Transaction (SET) and Secure Socket Layer (SSL) protocols.

In the opinion of the authors, highly efficient hardware implementations of cryptographic algorithms are provided in this book. However, these are not hard-fast solutions. The aim of the book is to provide an excellent guide to the design and development process involved in the translation from encryption algorithm to silicon chip implementation.

Preface

Overview of the Chapters

Chapter 1 provides background theory on both classical and modern cryptographic techniques. Examples of classical substitution and transposition ciphers are given and modern public and private key algorithms, hash functions and digital signatures are discussed. Typical cryptanalytic attacks are described. The motivation for implementing cryptographic algorithms in hardware is also outlined. The recent development of the Advanced Encryption Standard (AES) is reviewed. Each of the five AES finalists is compared in terms of structure, hardware implementation performance, advantages and disadvantages. The finite field mathematics in $GF(2^8)$ employed by the AES winner, Rijndael, is also explained in this chapter.

Perhaps the best known encryption algorithm, the Data Encryption Standard is introduced in *chapter 2*. A detailed description of the DES symmetric key algorithm is provided. Previous work on hardware-based DES designs is reviewed. This chapter describes a novel generic parameterisable DES architecture from which designs can be easily created for a range of specifications, such as scalability, single or Triple-DES functionality and modes of operation. A new key scheduling technique is also discussed, which, in a pipelined design, allows for the loading of different keys every clock cycle. The method is presented in relation to DES but can be applied to any pipelinable private key encryption algorithm.

In *chapter 3* the Rijndael private key algorithm is described. A review of existing Rijndael hardware designs is also provided. Two hardware-based fully pipelined Rijndael chip designs are presented which were among the first published implementations since the selection of Rijndael as the AES. The first is a novel generic parameterisable encryption-only architecture, which can generate designs to support each of the three key lengths required by the standard. In the second pipelined architecture the similarities between encryption and decryption are examined and exploited, so that a high-throughput design is achieved, while avoiding excess memory utilisation. A comparison between these new architectures and existing work is provided.

The Rijndael systems of chapter 3 are developed further in *chapter 4*. A Look-Up Table (LUT) based methodology is considered, which can achieve very high-speed AES designs, albeit at the expense of silicon area. To illustrate the technique, 128-bit key Rijndael iterative and pipelined encryptor designs are implemented in which the complex operations of the algorithm are replaced by LUTs. This chapter also presents a novel generic and migratable Rijndael architecture. Cores generated from this support each of the three AES key lengths, both encryption and decryption and four modes of operation. A hardware design that involves on-the-fly generation of sub-keys during decryption is also described. Once again, a comparison is provided between these new architectures and existing design work.

In *chapter 5*, hash algorithms are studied and security applications, which incorporate both hash and symmetric-key encryption algorithms, are considered. The SHA-1, SHA-384, SHA-512 and HMAC-SHA-1 hash algorithms are described in detail. One of the first hardware designs of the SHA-384 and SHA-512 algorithms is outlined. The Internet Protocol Security (IPSec) standard, which requires authentication and encryption, is described in the chapter. A novel single-chip IPSec architecture is presented, which comprises the generic Rijndael design outlined in chapter 4 and a HMAC-SHA-1 design. Other applications in which the IPSec architecture can be utilised are also examined.

The book finishes with a concluding summary and discusses directions for possible future research.

Terminology

Cryptography: The art or science of disguising messages.

Plaintext: The original message – also known as cleartext.

Encryption: The process of encoding or disguising a message to hide its contents.

Ciphertext: The encrypted or disguised message.

Decryption: The process of retrieving the original message from the ciphertext.

Key: A key word, number or electronic code that is used in the encryption and decryption process.

Cipher/ Cryptosystem: A system which performs encryption and decryption.

Cryptanalysis: The art or science of breaking cryptosystems.

Cryptology: The study of both cryptography and cryptanalysis.

Contents

1	**BACKGROUND THEORY**	**1**
	1.1. Introduction	1
	1.2. Cryptographic Algorithms	2
	1.3. Cryptanalysis	8
	1.4. Hardware-Based Cryptographic Implementation	10
	1.5. AES Development Effort	14
	1.6. Rijndael Algorithm Finite Field Mathematics	21
	1.7. Conclusions	26
2	**DES ALGORITHM ARCHITECTURES AND IMPLEMENTATIONS**	**28**
	2.1. Introduction	28
	2.2. DES Algorithm Description	29
	2.3. DES Modes of Operation	32
	2.4. Triple-DES	37
	2.5. Review of Previous Work	38
	2.6. Generic Parameterisable DES IP Architecture Design	40
	2.7. Novel Key Scheduling Method	48
	2.8. Conclusions	53
3	**RIJNDAEL ARCHITECTURES AND IMPLEMENTATIONS**	**57**
	3.1. Introduction	57
	3.2. Rijndael Algorithm Description	58

	3.3. Review of Rijndael Hardware Implementations	63
	3.4. Design of High Speed Rijndael Encryptor Core	65
	3.5. Encryptor/Decryptor Core	70
	3.6. Performance Results	72
	3.7. Conclusions	74

4 FURTHER RIJNDAEL ALGORITHM ARCHITECTURES AND IMPLEMENTATIONS — 77

- 4.1. Introduction — 77
- 4.2. Look-Up Table Based Rijndael Architecture — 78
- 4.3. Rijndael Modes of Operation — 84
- 4.4. Overall Generic AES Architecture — 87
- 4.5. Conclusions — 97

5 HASH ALGORITHMS AND SECURITY APPLICATIONS — 99

- 5.1. Introduction — 99
- 5.2. Internet Protocol Security (IPSec) — 100
- 5.3. IPSec Authentication Algorithms — 105
- 5.4. IPSec Cryptographic Processor Design — 107
- 5.5. Performance Results — 112
- 5.6. IPSec Cryptographic Processor Use in Other Applications — 115
- 5.7. SHA-384/SHA-512 Processor — 117
- 5.8. Conclusions — 121

6 CONCLUDING SUMMARY AND FUTURE WORK — 125

- 6.1. Concluding Summary — 125
- 6.2. Future work — 128

Appendix A – Modulo Arithmetic — 131

Appendix B – DES Algorithm Permutations and S-Boxes — 135

Appendix C – LUTs Utilised in Rijndael Algorithm — 139

Appendix D - LUTs in LUT-Based Rijndael Architecture — 143

Appendix E – SHA-384/SHA-512 Constants — 151

References — 153

Index — 159

Chapter 1

BACKGROUND THEORY

1.1. Introduction

The word 'cryptography' is derived from the greek words *kryptos*, which means hidden and *graphia* which means writing. Cryptography is the art of keeping secret the contents of a message transmitted over an unsecured communication channel. For example, the sender encrypts a message and thus, transforms its contents into an unintelligible form. The encrypted message or ciphertext is then transmitted over an unsecured channel. The receiver must decrypt the ciphertext to obtain the original message by performing an inverse transformation. The secrecy and security of the system relies on only the recipient having knowledge of the decryption transformation.

In the 1970s, Diffie and Hellman predicted that *'the development of computer controlled communication networks'* would provide *'effortless and inexpensive contact between people or computers on opposite sides of the world, replacing most mail and many excursions, with telecommunications'* [1]. In the last decade, their prediction has become a reality with the emergence of email, e-commerce, Virtual Private Networks (VPNs) and wireless communication technology. Private documents, which in the past, would have been hand-delivered and kept under lock and key are now typically created, sent and stored electronically [2]. However, this electronic age has led to an increased risk to information security and privacy and thus, the need for encryption. Traditionally, cryptography was only used by the military and diplomatic services for secure communication. However, today it is one of the principal tools employed to maintain privacy and confidentiality [3].

The history of cryptographic techniques from classical encryption to modern day algorithms is described in this chapter. Well-known attacks performed on cryptosystems are outlined. The advantages of hardware-based cryptographic implementations are also discussed. The development of the most recently designed cryptographic algorithm, the Advanced Encryption Standard (AES), is explained. Advantages and disadvantages of each of the five AES finalists, MARS, Rijndael, RC6, Twofish and Serpent are provided. Finally, the finite field mathematics utilised in the AES winner, Rijndael, is described.

1.2. Cryptographic Algorithms

1.2.1. Classical Encryption

One of the earliest accounts of secret writing dates back to the conflicts between Greece and Persia in the fifth century BC [4]. In 480 BC, Xerxes, leader of the Persians planned to attack the Athenian naval fleet stationed in the Bay of Salamis. The Persian fleet greatly outnumbered that of the Greeks. However, Demaratus, a Greek living in exile in a Persian city, decided to warn the Greeks. He scraped the wax from a wooden tablet, wrote of the planned assault and then reapplied the layer of wax, thus, hiding the message. When the tablet reached the Greeks, the layer of wax was scraped off and the message retrieved. Unknown to Xerxes, the Greeks were prepared for his attack. They enticed the Persian fleet into the Bay of Salamis, surrounded them and thwarted the assault.

Many famous historical figures such as Julius Caesar, Francis Bacon, Louis XIV, Cardinal Mazarin and Napoleon invented their own ciphers and used them for their secret correspondence. During the American War of Independence George Washington's spies employed a code system in which words were replaced with numbers from a code book. At the turn of the century, the French military used a ciphering machine known as Bazeries cylinder (named after the officer who invented it in 1891). Perhaps the most famous of all cryptography systems was the *Enigma* machine developed by the Germans during World War II. The success of Alan Turing and other code-breakers in cracking this cryptosystem had a significant bearing on the outcome of the war.

Substitution Cipher

Prior to the advent of computers encryption algorithms were primarily based on character substitution or transposition [5]. Julius Caesar was one of the first to use substitution ciphers to communicate in both domestic and military affairs [6]. In a substitution cipher, each letter in the plaintext is replaced with a different letter to form the ciphertext. The cipher used by Julius Caesar, known as the Caesar cipher, replaced every letter in a message with the letter 3 places to the right of it in the alphabet. For example, the letter *A* would be replaced by *D* and the letter *B* by *E*. An example of the Caesar cipher is given in Table 1-1.

Table 1-1. Example of the Caesar Cipher

Plaintext	C R Y P T O G R A P H Y
Ciphertext	F U B S W R J U D S K B

Polyalphabetic substitution ciphers use multiple alphabets to conceal the contents of a message. A well-known example is the Vigenère cipher. The Vigenère cipher uses 26 distinct alphabets. The first alphabet comprises all the letters shifted one place to the left. In the second alphabet, the letters are shifted two places and similarly for the remaining 24 alphabets. The alphabets are used in conjunction with a keyword, which is written above the message and repeated as required depending on the message length. Each letter in the keyword determines which alphabet is used to encrypt the corresponding message letter, as illustrated in Table 1-2. The Vigenère alphabets used in the example are given in Table 1-3.

Table 1-2. Example of the Vigenère Cipher

Keyword	S E C R E T S E C R E T
Plaintext	C R Y P T O G R A P H Y
Ciphertext	U V A G X H Y V C G L R

Table 1-3. Alphabets Utilised in Vigenère Cipher Example

	A	B	C	D	E	F	G	H	I	J	K	L	M	N	O	P	Q	R	S	T	U	V	W	X	Y	Z
C	C	D	E	F	G	H	I	J	K	L	M	N	O	P	Q	R	S	T	U	V	W	X	Y	Z	*A*	B
E	E	F	G	H	I	J	K	*L*	M	N	O	P	Q	R	S	T	U	*V*	W	*X*	Y	Z	A	B	C	D
R	R	S	T	U	V	W	X	Y	Z	A	B	C	D	E	F	*G*	H	I	J	K	L	M	N	O	P	Q
S	S	T	*U*	V	W	X	*Y*	Z	A	B	C	D	E	F	G	H	I	J	K	L	M	N	O	P	Q	R
T	T	U	V	W	X	Y	Z	A	B	C	D	E	F	G	*H*	I	J	K	L	M	N	O	P	Q	*R*	S

Transposition Cipher

In transposition ciphers, also known as permutation ciphers, the characters in a message are rearranged or transposed. In a columnar transposition cipher, the plaintext is inserted into a matrix in rows and the ciphertext is deduced by writing the letters down from each column. An example of a columnar transposition cipher is provided in Table 1-4.

Table 1-4. Example of Columnar Transposition Cipher

Plaintext	S E C U R E I N F O R M A T I O N
Ciphertext	S E R O E I M N C N A U F T R O I

S	E	C	U	R
E	I	N	F	O
R	M	A	T	I
O	N			

Some transposition ciphers also use key words. The number of letters in the keyword determines the number of columns, while the alphabetical priority of the letters is used to determine the order in which the columns are written to form the ciphertext, as shown in Table 1-5.

Table 1-5. Example of a Transposition Cipher Using a Keyword

Plaintext	S E C U R E I N F O R M A T I O N
Keyword	*S E C R E T*
Alphabetical Priority	*5 2 1 4 3 6*
	S E C U R E
	I N F O R M
	A T I O N
Ciphertext	C F I E N T R R N U O O S I A E M

Mechanical Encryption Devices

Cryptanalysis of classical ciphers, such as the substitution and transposition ciphers, is made possible because of redundancy in the linguistic structure of natural languages [7]. Consequently more sophisticated systems were developed and mechanical encryption devices were invented in the 1920s to facilitate the use of these cryptosystems [5]. Rotor machines, which resembled typewriters, provided arbitrary permutations of the alphabet. These consisted of a number of rotors, with each rotor having 26 positions. The rotor rotated each time a letter was entered into the machine. This altered the substitution pattern, making the rotor machine equivalent to a polyalphabetic cipher. However, the machine used multiple rotors, with the output position of one connected to the input position of another, thus, providing far greater complexity. For example, in a 3-rotor machine, the first rotor might substitute an *A* with *E*, the second rotor might substitute an *E* with *P* and the third rotor might substitute a *P* with *K*. Therefore the initial plaintext letter, *A* would be represented as a *K* in the ciphertext. The German *Enigma* and the Japanese *Purple* machines, used during World War II, are the most famous rotor machines.

1.2.2. Modern Encryption

Modern encryption techniques are based on sophisticated and complex mathematics. Whereas classical ciphers depended on the security of the entire encryption process, in more modern cryptosystems, the encryption algorithm can be revealed without compromising security. In these ciphers a key is used along with the algorithm and thus, the security of the system relies entirely on the secrecy of this key. The number of possible keys is so large that it proves infeasible for an attacker to attempt to uncover the individual key that corresponds to a message. The complex mathematics used in the algorithms makes it difficult to determine the key from the publicly available information. Modern cryptographic algorithms include public and private key cryptosystems, hash algorithms and digital signatures.

Public Key Algorithms

In a *public key* or *asymmetric* cryptosystem, one key, known as the public key, is used to encrypt the message and a second key, known as the private key is used to decrypt it, as illustrated in Figure 1-1. In 1976, Diffie and Hellman were the first to propose a public key cryptosystem [8]. In order to communicate securely, the sender and recipient of a message must both have two keys – a public and a private key. Their public keys are made generally available, while the private keys are kept secret and are only known by the individual owner. To transfer a message, the sender encrypts it using the recipient's public key. The resulting ciphertext can then be sent along an unsecured channel. The receiver decrypts the ciphertext using the private key.

Figure 1-1. Public Key Cryptosystem

It is computationally infeasible to determine the decryption key from the publicly available cryptographic algorithm and encryption key [9]. Public key algorithms are typically based on mathematical problems such as the difficulty in factoring large prime numbers or in computing discrete logarithms in a finite field.

One of the most commonly used public key algorithms is RSA, which is named after its designers Ronald Rivest, Adi Shamir and Leonard Adleman. It is based on the non-deterministic polynomial-time (NP) problem, the prime factorisation of a large number. NP problems require a long period of time to be solved, even though a proposed solution can be easily verified. As the problem increases, the number of computational steps needed to find a solution increases exponentially, while the number of steps required to check a possible solution increases only in proportion to a polynomial function. Other public key cryptosystems include the Diffie-Hellman algorithm, ElGamal and McEliece.

Private Key Algorithms

In a *private key* or *symmetric* algorithm, the same key is used to encrypt and decrypt a message. An outline of a private key cryptosystem is given in Figure 1-2. Since only one key is used, the security of the system relies entirely on the secrecy of the key. The key must be transferred from the sender to the receiver of a message via a known secure channel.

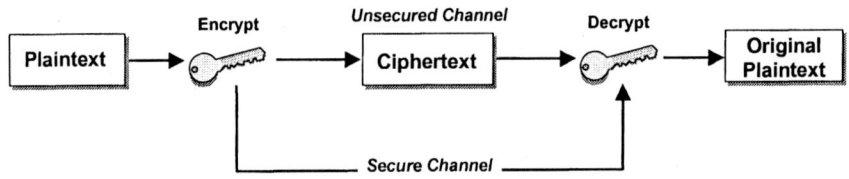

Figure 1-2. Private Key Cryptosystem

There are two classes of symmetric key algorithms, stream ciphers and block ciphers. Stream ciphers encrypt single bits of plaintext, while block ciphers operate on a fixed size data block. A famous example of a stream cipher is the one-time pad, which was designed by Gilbert Vernam in 1917 [10], and hence, it is also known as the Vernam cipher. The one-time pad uses a string of completely random bits, called the keystream. The keystream must be the same length as the message and is bitwise XORed with the plaintext to produce the ciphertext. In terms of alphabetical letters, encryption is the addition modulo 26 of a plaintext character with the corresponding keystream character [5]. Since each keystream is truly random and used only once, the one-time pad is a perfectly secure encryption system. The main disadvantage to using the cipher is that truly random keys of the same length as the message must be generated and securely transmitted to the recipient. There are very few situations where this is practical [11]. An example of the Vernam cipher is depicted in Table 1-6.

Table 1-6. Example of the Vernam One-Time Pad

Plaintext	P E R F E C T C I P H E R
Keystream	L Q T A L M C I U P F H Z
Ciphertext	B V L G Q P W L D F N M R

Block ciphers are the most common form of private key cryptosystems. Perhaps the most famous of all encryption algorithms is the DES block cipher, which is discussed in chapter 2. Typically, block ciphers have a *feistel* structure, which was designed by Horst Feistel in the 1970s [12]. A feistel cipher involves multiple iterations of a simple non-linear function. One half of the plaintext is operated on by the function, the result of which is XORed to the other half. The two halves are then swapped prior to the next iteration. Examples of feistel algorithms include, DES, LOKI, MARS, Serpent and Twofish. Other well-known block ciphers are IDEA, RC6 and the recently developed Rijndael algorithm, which is studied in chapters 3 and 4.

Hash Functions

A hash function transforms a variable-length message into a much shorter fixed-length output. Hash algorithms are used to ensure the integrity of a message. A simple example is to XOR each byte of input data to produce a one-byte hash result. However, in order for a hash algorithm to be cryptographically useful, it must be a one-way hash function [6]. A one-way function is a function that is relatively easy to compute but computationally infeasible to reverse.

A message authentication code (MAC) is a hash function, which utilises a key. The sender generates a hash of the message using a secret key and sends this hash value along with the encrypted message. Similar to private key algorithms, it is necessary to send the key via a known secure channel. The receiver decrypts the ciphertext and using the same secret hash key can create a hash of the received message. If the hash values correspond then the receiver is assured that the message has not been interfered with during transit and that its sender is authentic. Thus, when a hash function is utilised as a MAC, it provides authentication in addition to ensuring message integrity. Private key algorithms such as DES and Rijndael can be used to generate MACs. Other examples of hash functions include RIPEMD-160, MD5 and the Secure Hash Algorithm (SHA), which is discussed further in chapter 5.

Digital Signatures

Digital signature algorithms are utilised to sign and authenticate documents. A digital signature provides the same effect as a real signature in that it verifies that a message originates from a specific person. It consists of two components, a signing algorithm and a verifying algorithm. The sender encrypts a message with the recipient's public key and using a signing algorithm, encrypts the signature with his own private key. On receiving the encrypted message, the recipient can authenticate the message by decrypting the signature with a verifying algorithm and the sender's public key. He can then decrypt the ciphertext with his own private key.

Many public key ciphers are used for digital signatures such as RSA and ElGamal. However, in 1991, the NIST proposed the Digital Signature Standard (DSS), which comprises the SHA hash function and a Digital Signature Algorithm (DSA). The DSA algorithm cannot be used to encrypt or decrypt data and is solely used to create signatures [13]. The algorithm is based on the difficulty of computing discrete logarithms and is an adaptation of the ElGamal public key algorithm. The SHA hash algorithm is utilised to create a hash of the message. The resulting hash is input into the signature algorithm to produce the signature.

Advantages and Disadvantages of Public and Private Key Algorithms

The most common modern encryption techniques are public and private key cryptosystems. Private key algorithms are used for bulk data encryption and are at least 1000 times faster than public key cryptosystems [5]. However, private key algorithms require the secure generation and distribution of the secret key. Also, it is necessary to change the key frequently to avoid the risk of it being compromised. Conversely, public key algorithms are generally used in key distribution and not in the encryption of messages since they are limited by computational cost and thus, are slow in speed. A primary advantage of public key ciphers is their use in digital signatures.

Many applications, however, exploit the advantages of both public and private key algorithms. The private key algorithm is used to encrypt the message with a symmetric key. This symmetric key is then encrypted using a public key cipher and the recipient's public key. The encrypted message and encrypted key are then sent along an unsecured channel to the receiver of the message. The recipient decrypts the encrypted key using a private key and can then decrypt the message using the decrypted symmetric key to obtain the original message.

1.3. Cryptanalysis

Cryptanalysis can be described as the art of breaking cryptosystems. To break a cipher involves finding a weakness in the cipher that can be exploited with a complexity less than a brute-force attack [14]. A brute-force attack simply involves trying all possible keys in a sequence. Whether this is feasible or not depends on the size of the key. A fundamental assumption of cryptanalysis is *Kerkhoff's principle*, which states that the security of a cryptosystem must not depend on keeping the cryptographic algorithm

Background Theory

secret, but only on keeping secret the key [4]. Successful cryptanalysis involves a combination of mathematics, inquisitiveness, intuition, persistence and powerful computing resources.

The two main classes of attacks are passive and active attacks. A passive attack is where communication is monitored and thus, the confidentiality of information is threatened. An active attack, on the other hand, is where the attacker attempts to modify the information and so the confidentiality, integrity and authenticity of a message are compromised. The objective in all these attacks is to determine the key [10].

The most common passive attacks include:

Ciphertext-Only
An attacker has access to a block of ciphertext only.

Known-Plaintext
An attacker has access to both a block of plaintext and the corresponding ciphertext.

Chosen-Plaintext
An attacker has gained access to the encryption process of a cryptosystem and therefore, has the ability to input plaintext and construct the corresponding ciphertext.

Chosen-Ciphertext
This is the reverse of a chosen plaintext attack. The attacker has access to the decryption process of a cryptosystem and is able to input ciphertext and reconstruct the original plaintext.

Three well-known active attacks are:

Man-in-the-Middle (MITM)
This is where an attacker intercepts a communication channel between two parties. The attacker can then retrieve information and send on altered messages without the knowledge of either party. Keys are easily compromised in a MITM attack. Typically, hash functions are used to thwart such attacks.

Timing
Cryptographic algorithms vary in the time it takes to process different data and key inputs. By carefully measuring the amount of time required to perform certain operations, information can be retrieved and indeed, secret keys can even be uncovered [15].

Power Analysis
This is a recently discovered technique that involves interpreting power consumption measurements of various cryptographic operations to retrieve information [16]. Features such as DES permutation and shift operations can be easily differentiated, as their power consumption is visibly different. Differential Power Analysis (DPA) is an even more powerful method of attack, in which statistical analysis and error correction techniques are also used to deduce information.

The science of cryptology is continually driven forward by the constant battle between cryptographers trying to secure information and cryptanalysts attempting to break cryptosystems.

1.4. Hardware-Based Cryptographic Implementation

1.4.1. Encryption Design Architectures

A number of different architectures can be considered when designing encryption algorithms [17]. These are described as follows:

Iterative Looping (IL)
Only one round is designed, hence for an n-round algorithm, n iterations of that round are carried out to perform an encryption

Loop Unrolling (LU)
Involves the unrolling of multiple rounds

Pipelining (P)
Achieved by replicating a round function and placing registers between each round to control the flow of data

Sub-Pipelining (SP)
The addition of further registers into a pipelined design when a round function of the pipelined architecture is complex. It decreases the pipeline's delay between stages but increases the number of clock cycles required to perform an encryption

A pipelined architecture will provide the highest overall throughput. Thus, if a high-speed design is required, a fully pipelined architecture should be utilised. Further improvements in speed can be achieved by sub-pipelining the design. However, this incurs additional delays in the output data. If, on

Background Theory

the other hand, area is a consideration, an iterative architecture will produce the most compact design. For specific speed and area requirements, hybrid architectures can be employed.

1.4.2. Advantages of Hardware-Based Implementation

Traditionally, Digital Signal Processing (DSP) algorithms have been implemented in software on enhanced and specifically designed microprocessors. Although this is a low cost solution, particularly for high product volumes, the performance often fails to meet the requirements of many applications. Alternative solutions, which are currently being utilised, include implementation on arrays of microprocessors, fixed function processors and custom hardware solutions such as Application Specific Integrated Circuits (ASICs). In the past decade, however, there has been a major increase in the use of Programmable Logic Devices (PLDs), such as Field-Programmable Gate Arrays (FPGAs) and Complex Programmable Logic Devices (CPLDs), for the implementation of high-level DSP functions [18]. In particular, FPGAs offer many advantages over both ASIC and software solutions.

Hardware-based implementations are inherently faster than software implementations. However, like microprocessors, FPGAs offer in-circuit reprogrammability. In addition, they can also support high levels of parallelism, which maximises throughput and allows for the design of fully pipelined and functionally tailored architectures. Since FPGAs have a register-rich architecture, pipelining requires no additional resources [19]. Hardware designs of encryption algorithms prove much faster than equivalent software architectures (typically several orders of magnitude) and since there is an increasing need to perform encryption on data in real time, this is very important. Increased bandwidth requirements are a major factor pushing encryption out of software, since software cannot process encrypted data at gigabit speeds [20]. One possible solution is to install more servers. However, this is far more expensive than utilising dedicated hardware cryptographic systems. Encryption is a computationally intensive process that comprises the complex manipulation of data and general-purpose microprocessors cannot perform such operations efficiently. Also, hardware cryptographic systems are physically more secure than software systems. In software-based cryptosystems the algorithm keys are stored in the unprotected memory of a processor. If a symmetric key is compromised, the encrypted data is no longer secure and can be easily read. If an asymmetric private key is accessed, it can be used to intercept and modify messages and to generate false digital signatures, which would be attributed to the key's rightful owner. In hardware-based encryption the keys can be stored in tamper-resistant devices such as smart cards, thus, ensuring their integrity and authenticity [21].

Although slower than ASIC devices, FPGAs share the cost advantages of high production volumes and none of the Non-Recurrent Engineering (NRE) costs or fabrication delays associated with ASIC development. Also, when using FPGAs, the long ASIC design cycle is almost eliminated since there are no delays for prototype development and design revisions are easily implemented. In the past, the low gate densities and high costs per unit of FPGA devices relegated them to small, low-volume designs. However, within the last number of years, the increase in performance and density of FPGAs and features, such as embedded microprocessor and memory cores, have led to these devices becoming viable and highly attractive alternatives to ASICs in many applications. Ten years ago, the size of an average ASIC device was 10,000 gates, while the largest FPGA had one tenth of that capacity. Today, the average ASIC contains 250,000 gates and the largest FPGA comprises double that capacity [22]. In cryptographic applications, FPGAs provide considerable flexibility over ASIC devices particularly for private key encryption algorithms, which seem to fit extremely well with the characteristics of an FPGA [23]. It is possible to reconfigure an FPGA to switch between cryptographic algorithms and indeed between encryption and decryption modes.

1.4.3. FPGA Structure

In this research, the new generic cryptographic architectures described have been implemented on Xilinx Virtex, Virtex-E or Virtex-II FPGA technology [24] for demonstration purposes. However, as will be discussed, these architectures and designs are easily migratable to other silicon technologies such as ASICs and CPLDs.

Xilinx Virtex FPGAs are high-density devices and comprise between 50,000 to 1 million system gates depending on the particular device used. They have been developed using a 0.22µm 5-layer metal CMOS process. A Virtex device is made up of an array of Configurable Logic Blocks (CLBs), which are surrounded by Input/Output Blocks (IOBs) and interconnected by versatile routing resources, as illustrated in Figure 1-3. [24]. A CLB comprises two slices, each containing two Look-Up Tables (LUTs) and two flip-flops. A LUT can be configured as a 16 x 1-bit Random Access Memory (RAM) or combined with LUTs in other slices to created larger RAMs. An IOB has three storage elements that have independent clock enable signals. A Virtex FPGA also contains two columns of large embedded memory blocks, known as Block RAM (BRAM). The two columns are located on the vertical edges of a device. A BRAM is a synchronous 4096-bit memory block, which can be configured to support varying address and data widths.

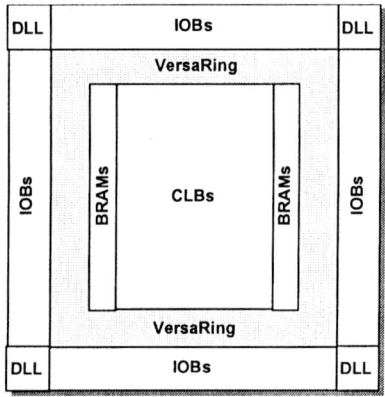

Figure 1-3. Virtex Architecture Overview

The Xilinx Virtex-E family of FPGAs consists of fast, high-density devices with 58,000 to 4 million system gates. They are designed for low-power operation and have been fabricated using a 0.18 µm 6-layer metal CMOS process. Similar to Virtex FPGAs, Virtex-E devices comprise CLBs, IOBs and BRAMs. However, the BRAMs are arranged in multiple columns throughout the device. The Virtex-E Extended Memory (EM) family of FPGAs are an extension of the Virtex-E family, and comprise additional BRAM. The XCV812E extended memory device contains the highest number of memory blocks with 280 BRAMs.

Virtex-II FPGAs range from 40,000 to 10 million system gates in density. They have been developed using a 0.15µm 8-layer metal CMOS process. The architecture is optimised for high-speed and low power consumption. It includes CLBs, IOBs and 2 to 6 columns of BRAMs. A CLB in a Virtex-II device contains four slices. Each BRAM is a synchronous 18 Kbit memory block that can be configured to form single or dual port RAMs of various depths and widths. Virtex-II devices also contain 18-bit x 18-bit multipliers, which are located in columns adjacent to each BRAM column.

1.5. AES Development Effort

1.5.1. DES Algorithm Downfall

In January 1997, the RSA Data Security company issued a challenge to break the US government's DES algorithm. In June of that year, the challenge was solved by the DESCHALL team who successfully recovered the 56-bit DES key. The DESCHALL team, led by Rocke Verser, Matt Curtin and Justin Dolske, adopted a brute-force attack in their attempt to uncover the 56-bit DES key. DES has 2^{56} or approximately 10^{17} possible keys. A computation with such a magnitude of operations was unlikely for most computer users in the mid-1970s. However, today, massively parallel machines can threaten the security of DES using a brute-force attack. DESCHALL entailed a large-scale distributed computing project on the Internet [25]. The project linked together thousands of volunteers, each checking different keys. By writing programs for Unix, Macintosh, Windows and OS/2 operating systems, the DESCHALL team utilised the computing power of large workstations as well as personal home computers. In the end, the key was discovered after only searching 24.6% of all possible keys [26]. Recently, several brute-force recoveries of 56-bit DES keys have been demonstrated. In 1999, the Electronic Frontier Foundation (EFF) built a key search machine, which can find a DES key in less than 23 hours [27].

1.5.2. AES Development

In 1997, the National Institute of Standards and Technology (NIST) requested candidates for a new Advanced Encryption Standard (AES) algorithm to replace DES, realising that the algorithm's 56-bit key was no longer sufficient to provide the necessary security in many applications. As an interim measure they adopted and standardised Triple-DES, which uses three passes of the DES algorithm and a 112 or 168-bit key. The NIST required an algorithm, which would provide good security for the foreseeable future, be efficient and suitable for various platforms and environments, and provide flexibility to accommodate future requirements [28].

In August 1998, at the first AES conference, the NIST began round 1 of technical analysis for the AES development effort by announcing 15 candidate algorithms. The second AES conference was held in March 1999. At this conference the technical analysis was presented and discussed, along with views as to which candidates should be selected as finalists for round 2.

Background Theory

The following five AES finalists were announced in August 1999:
- *MARS* : developed by the IBM Corporation, America
- *RC6* : developed by RSA Laboratories, America
- *Rijndael* : developed by Joan Daemen and Vincent Rijmen of the Katholieke University Leuven, Belgium
- *Serpent*: developed by Ross Anderson, Eli Biham and Lars Knudsen of the United Kingdom, Israel and Norway respectively
- *Twofish*: developed by Bruce Schneier, John Kelsey, Doug Whiting, David Wagner, Chris Hall and Niels Ferguson of Couterpane Systems, America

1.5.3. Comparison of the AES Finalists

NIST specified that proposed AES algorithms must implement a symmetric block cipher capable of supporting a data block size of 128-bits and keys of 128, 192 and 256-bits in length. They wanted an algorithm with security at least as effective as Triple-DES, but with significantly improved efficiency.

Structure

A number of different algorithm structures exist [29]:

A *product cipher* combines two or more encryption operations whose combination is more secure than the individual components. Typical operations include transpositions, substitutions, linear transformations, XOR and other arithmetic functions.

A *substitution-permutation* (SP) network is a product cipher which comprises alternating stages of substitutions and permutations.

An *iterated cipher* is one which involves an encryption process that has several iterations or rounds. In each round the same transformation or round function is applied to the data utilising a sub-key, which is derived from the cipher key.

A *feistel* cipher, as described earlier, is an iterated algorithm that maps a $2t$-bit plaintext, (L_0, R_0), for t-bit blocks L_0 and R_0, to a ciphertext (R_r, L_r), through an r-round process where $r \geq 1$.

- **Twofish**

    The Twofish algorithm has a feistel network structure and contains 16 rounds.

- **Rijndael, Serpent**

    Rijndael and Serpent are both substitution-permutation (SP) algorithms. Rijndael utilises 10, 12 and 14 rounds for 128-bit, 192-bit and 256-bit keys respectively. The Serpent algorithm consists of 32 rounds.

- **MARS, RC6**

  Both the MARS and RC6 algorithms have modified feistel structures. MARS consists of 32 rounds. The initial eight rounds are unkeyed, the next sixteen are keyed rounds and the final eight are unkeyed rounds. The RC6 algorithm contains 20 rounds. The key schedule of the RC6 algorithm is identical to that of the RC5 algorithm.

Advantages and Disadvantages

Numerous studies on each of the five AES finalists were presented at the third AES conference (AES3) held in April 2000 [30]. From these studies a summary of the advantages and disadvantages of each algorithm was compiled as follows:

Twofish Algorithm

Advantages:
- High security margin – reasonable complexity
- Twofish round function has proven to be the strongest round function of any of the finalists
- Suitable for restricted-space environments
- Compact hardware implementations are possible
- Supports arbitrary key sizes up to 256-bits

Disadvantages:
- Difficult to defend against timing and power attacks
- Average performance when implemented in software
- Key dependent S-Boxes

MARS Algorithm

Advantages:
- Has one of the highest security margins, both in terms of number of rounds and in terms of diversity
- Can support key sizes from 128 bits to 448 bits

Disadvantages:
- Difficult to implement in memory constrained environments
- Complex
- Average encryption/decryption performance when implemented in software
- Difficult to defend against power and timing attacks
- In hardware implementations, MARS has above average area requirements and below average throughput
- Not suited to key-agile systems (e.g. IPSec)

RC6 Algorithm
Advantages:
- Simple, elegant round function
- One of the fastest algorithms when implemented in C-language in the majority of software studies carried out on 32-bit platforms
- RC5 has been in existence for almost six years - the key schedule of the RC6 algorithm is identical to that of the RC5 algorithm - existing analysis on RC6 is not only by far the most extensive of any of the finalists, it is also the most accurate and the most detailed
- Suitable for restricted-space environments
- Can be implemented compactly in hardware
- Can support variable key, block and round sizes

Disadvantages:
- Round function possibly too simple – does not use S-Boxes
- Only has an adequate security margin – algorithms are required to last at least 20 years
- Difficult to defend against power and timing attacks

Rijndael Algorithm
Advantages:
- In hardware implementations, offers the highest throughput for feedback modes and the second highest for non-feedback modes of operation
- Performs well in software implementations – the key setup time is the fastest of all the finalists
- Easy to defend against power and timing attacks
- Very well suited for memory constrained environments
- It is easily implemented on a wide range of platforms
- Supports 128-bit, 192-bit and 256-bit key and data block lengths

Disadvantages:
- Adequate security margin

Serpent Algorithm
Advantages:
- Simple structure
- Very high security margins in terms of number of rounds – very strong mixing – most secure of all the algorithms
- Fastest of the AES finalists when implemented in hardware for non-feedback modes and second highest in feedback modes of operation
- Relatively easy to defend against power and timing attacks
- Suitable for memory constrained environments
- Can support key sizes up to 256-bits

Disadvantages:
- Generally the slowest of the finalists when implemented in software

Hardware Implementation Performance Evaluation

In particular, studies were carried out on the hardware implementation of the five finalist algorithms. These were also presented at the third and final AES conference. A summary of the principal hardware-based studies is provided below.

Implementations on FPGA Devices

Dandalis, Prasanna and Rolim [23] carried out implementations of the five finalists on the Xilinx Virtex family of FPGAs. Iterative designs of each algorithm are implemented with only feedback modes of operation being considered. Table 1-7 summarises the results obtained showing the throughput achieved and the area utilised by each algorithm.

The Rijndael algorithm has the highest encryption rate and also achieves a very efficient hardware design. Serpent achieves the most compact implementation with the lowest area utilisation.

Table 1-7. FPGA Implementation Results Obtained by Dandalis, Prasanna and Rolim [23]

Algorithm	Archit.	Throughput (Mbits/sec)	Area (CLB slices)	Throughput/Area (Mbps/slices)
RC6	IT	112.9	2650	0.04
Rijndael	IT	353	5673	0.06
Serpent	IT	149	2550	0.06
Twofish	IT	173	9363	0.02
MARS	IT	203.8	6896	0.03

Elbirt, Yip, Chetwynd and Paar [17] carried out their study on four of the five finalists. A summary of the results obtained is outlined in Table 1-8.

The MARS algorithm was excluded from the study on the basis that it would achieve a poor performance when compared to the other finalists due to its use of large S-Boxes and modulo 2^{32} multiplications. A number of different architectures were considered when designing the algorithms such as loop unrolling (LU), partial pipelining (PP) and sub-pipelining (SP). However, only the encryption function of each algorithm was implemented and the algorithms were compared with respect to throughput optimisation. The algorithms were implemented on XCV1000 Xilinx Virtex FPGAs.

Serpent outperforms the other finalists both in terms of throughput achieved and area utilised. It is evident that Serpent is well suited for an FPGA implementation from a performance perspective. All four algorithms

Background Theory

achieve Gigabit encryption rates, which is at least one order of magnitude faster than the best reported software realisations.

Table 1-8. FPGA Implementation Results Obtained by Elbirt, Yip, Chetwynd and Paar [17]

Algorithm		Archit.	Throughput (Mbits/sec)	Area (CLB slices)	Throughput/Area (Mbps/slices)
RC6:	Feedback	PP	126.5	3189	0.04
	Non-feedback	SP	2398	10856	0.22
Rijndael:	Feedback	LU	300	5302	0.06
	Non-feedback	SP	1938	10992	0.18
Serpent:	Feedback	LU	444	7964	0.06
	Non-feedback	PP	4860	9004	0.54
Twofish:	Feedback	SP	120	3053	0.04
	Non-feedback	SP	1585	9345	0.17

Gaj and Chodowiec [31] carried out FPGA implementations on each of the five algorithms using Xilinx Virtex XCV1000 FPGA devices. Iterative designs were implemented and therefore, only feedback modes were considered. The algorithm key schedules were not included in implementation. Table 1-9 outlines the results obtained.

Table 1-9. FPGA Implementation Results Obtained by Gaj and Chodowiec [31]

Algorithm	Archit.	Throughput (Mbits/sec)	Area (CLB slices)	Throughput/Area (Mbps/slices)
RC6	IT	103.9	1139	0.09
Rijndael	IT	331.5	2902	0.11
Serpent	IT	339.4	4438	0.08
Twofish	IT	177.3	1076	0.16
MARS	IT	39.8	2737	0.01

They classify the five algorithms depending on their performance characteristics. The first class includes Twofish and RC6, both of which are compact low-cost implementations with medium speed compared to other candidates. They are the only two algorithms of the five that can be implemented using low cost FPGA Xilinx XC4000 devices. The second class includes Serpent and Rijndael. Both guarantee very high speeds at the cost of the relatively large area. The third class contains MARS. It shows the worst hardware characteristics of the five candidates. It is the slowest algorithm, eight times slower than the fastest Serpent algorithm. It also utilises over twice the area used by the ciphers in the first group.

Implementations on ASIC Devices

ASIC implementations of the five AES finalists using a 0.5µm standard cell library were considered by Weeks, Bean, Rozylowicz and Ficke [32]. In this study each algorithm was implemented using both an iterative architecture and a pipelined architecture. The results obtained are illustrated in Table 1-10. The Twofish algorithm provides the smallest area of the iterative designs with Rijndael achieving the highest throughput. Of the pipelined architectures the Twofish algorithm has again the smallest area while Serpent achieves the fastest encryption rate. In both types of hardware implementation design, the MARS algorithm is the most inefficient.

Table 1-10. ASIC Implementation Results Obtained by Weeks, Bean, Rozylowicz, Ficke [32]

Algorithm	Archit.	Throughput (Mbits/sec)	Area (mm^2)	Throughput/ Area (Mbps/mm^2)
RC6: Pipelined	P	2197	453	4.85
Iterative	IT	102.8	19.2	5.35
Rijndael: Pipelined	P	5745	420	13.68
Iterative	IT	605.8	33.8	17.92
Serpent: Pipelined	P	8030	438.6	18.31
Iterative	IT	202.3	23.3	8.68
Twofish: Pipelined	P	2274	225.3	10.1
Iterative	IT	105	16	6.56
MARS: Pipelined	P	2189	1333	1.64
Iterative	IT	56.7	126.8	0.45

Ichikawa, Kasuya and Matsui [33] analysed the AES finalists using Mitsubishi Electric's CMOS 0.35µm ASIC design library. A summary of the results obtained is given in Table 1-11. Full loop unrolled designs of each algorithm were implemented and feedback modes of operation considered. Rijndael is the fastest algorithm among these implementations and is also the most efficient in terms of area utilisation. Serpent is the second fastest of the five algorithms.

In April 2000, at the third AES Candidate Conference, round 2 of the technical analysis was presented and discussed, along with views as to which of the finalists should be selected as the AES winner(s). In October 2000, the Rijndael algorithm was selected by the NIST as the new AES and in November 2001, it replaced DES as the Federal Information Processing

Background Theory   21

Encryption Standard. The NIST judged Rijndael to be the best overall algorithm in terms of security, performance, efficiency, flexibility and implementation characteristics.

Table 1-11. ASIC Implementation Results Obtained by Ichikawa, Kasuya and Matsui [33]

Algorithm	Archit.	Throughput (Mbits/sec)	Area (gatesx10^3)	Throughput/Area (Mbps/ gatesx10^3)
RC6	LU	204	1643	0.12
Rijndael	LU	1950	612.8	3.18
Serpent	LU	931.6	503.8	1.85
Twofish	LU	394	431.9	0.91
MARS	LU	225.6	2935.8	0.08

1.6. Rijndael Algorithm Finite Field Mathematics

The following section outlines the finite field mathematics in $GF(2^8)$ utilised in the Rijndael algorithm.

The elements of a finite field can be represented in several different ways. In the Rijndael specification, the classical polynomial representation is used. A byte, b: $b_7\ b_6\ b_5\ b_4\ b_3\ b_2\ b_1\ b_0$, is considered as a polynomial with coefficients in the finite field, $\{0,1\}$. The polynomial is represented as:

$$b_7 x^7 + b_6 x^6 + b_5 x^5 + b_4 x^4 + b_3 x^3 + b_2 x^2 + b_1 x + b_0 \qquad (1.1)$$

For example, the byte, 1000 1011, (0x8B in hexadecimal) corresponds to the polynomial,

$$x^7 + x^3 + x + 1 \qquad (1.2)$$

1.6.1. Addition

Utilising the polynomial representation, the addition of two values is the sum modulo 2 of the coefficients. In binary notation, this addition is a simple bitwise XOR.

For example,

$$\begin{aligned}&(x^7 + x^3 + x + 1) + (x^7 + x^6 + x^5 + x^3 + x^2 + x + 1)\\ &= x^6 + x^5 + x^2\end{aligned} \qquad (1.3)$$

and in binary,

$$1000\ 1011\ \text{XOR}\ 1110\ 1111 = 0110\ 0100 \qquad (1.4)$$

1.6.2. Multiplication

Multiplication in $GF(2^8)$ corresponds to multiplication of polynomials modulo an irreducible binary polynomial of degree 8. A polynomial is irreducible if it has no divisors other than itself and 1 [34].

In the Rijndael algorithm, this polynomial is $m(x) = 0x11B$, where

$$m(x) = x^8 + x^4 + x^3 + x + 1 \qquad (1.5)$$

Hence, the result of the multiplication will always be a polynomial of degree below 8. For example the multiplication of the two hexadecimal numbers, 0x63 and 0x15, using polynomial representation is,

$$(x^6 + x^5 + x + 1)(x^4 + x^2 + 1)$$

$$= x^{10} + x^9 + x^5 + x^4 + x^8 + x^7 + x^3 + x^2 + x^6 + x^5 + x + 1$$

$$= x^{10} + x^9 + x^8 + x^7 + x^6 + x^4 + x^3 + x^2 + x + 1 \qquad (1.6)$$

The result of $(x^{10} + x^9 + x^8 + x^7 + x^6 + x^4 + x^3 + x + 1)$ modulo $(x^8 + x^4 + x^3 + x + 1)$ is $x^7 + x^4 + x^3 + x^2 + x$ or 0x9E, as illustrated in Figure 1-4. (Remember that the polynomial coefficients lie in the finite field, {0,1})

$$\require{enclose}
\begin{array}{r}
x^2 + x + 1 \\[-3pt]
\end{array}$$

$$x^8 + x^4 + x^3 + x + 1 \enclose{longdiv}{x^{10} + x^9 + x^8 + x^7 + x^6 + x^4 + x^3 + x^2 + x + 1}$$

$$\begin{array}{r}
- (x^{10} + x^6 + x^5 + x^3 + x^2) \\ \hline
x^9 + x^8 + x^7 + x^5 + x^4 + x + 1 \\
- (x^9 + x^5 + x^4 + x^2 + x) \\ \hline
x^8 + x^7 + x^2 + 1 \\
- (x^8 + x^4 + x^3 + x + 1) \\ \hline
x^7 + x^4 + x^3 + x^2 + x
\end{array}$$

Figure 1-4. Modulo Division Example

1.6.3. Multiplication of Constants

Multiplication by x can be implemented at byte level as a *left shift* and a subsequent *conditional bitwise XOR* with the hexadecimal value 0x1B. This can be exploited to accommodate multiplication by any constant. For example, 0x63 * 0x15 can be calculated as follows:

- 0x63 * 0x02 = 0110 0011 * 0000 0010

  Shift 0x63 left by 1 place = 1100 0110 = 0xC6   (1.7)

- 0x63 * 0x04 = 0110 0011 * 0000 0100

  Shift 0xC6 left by 1 place = 1 1000 1100 (Result is 9-bits in length)

  => XOR with 0x1B = 1 1000 1100 ⊕ 0001 1011

  = 1001 0111 = 0x97   (1.8)

- 0x63 * 0x08 = 0110 0011 * 0000 1000

    Shift 0x97 left by 1 place =1 0010 1110   (Result is 9-bits in length)

    => XOR with 0x1B = 1 0010 1110 ⊕ 0001 1011

    = 0011 0101   = 0x35 　　　　　　　　　　　　　　　　(1.9)

- 0x63 * 0x10 = 0110 0011 * 0001 0000

    Shift 0x35 left by 1 place = 0110 1010   = 0x6A 　　　　(1.10)

Hence,

0x63 * 0x15 = 0x63 * (0x01 ⊕ 0x04 ⊕ 0x10)

= 0x63 ⊕ 0x97 ⊕ 0x6A

= *0110 0011* ⊕ *1001 0111* ⊕ *0110 1010*

= *1001 1110*   = 0x9E 　　　　　　　　　　　　　　　(1.11)

As is expected, the result obtained in equation (1.11) corresponds to the result achieved in Figure 1-4.

1.6.4. Multiplication of Two Polynomials

Rijndael uses the concept of 4-byte vectors to correspond with polynomials of degree 4. The product of two such polynomials, $a(x) = a_3 x^3 + a_2 x^2 + a_1 x + a_0$ and $b(x) = b_3 x^3 + b_2 x^2 + b_1 x + b_0$ is $c(x)$ where,

$$c(x) = c_6 x^6 + c_5 x^5 + c_4 x^4 + c_3 x^3 + c_2 x^2 + c_1 x + c_0 \tag{1.12}$$

with,
$$\begin{aligned}
c_0 &= a_0 * b_0 \\
c_1 &= a_1 * b_0 \oplus a_0 * b_1 \\
c_2 &= a_2 * b_0 \oplus a_1 * b_1 \oplus a_0 * b_2 \\
c_3 &= a_3 * b_0 \oplus a_2 * b_1 \oplus a_1 * b_2 \oplus a_0 * b_3 \\
c_4 &= a_3 * b_1 \oplus a_2 * b_2 \oplus a_1 * b_3 \\
c_5 &= a_3 * b_2 \oplus a_2 * b_3 \\
c_6 &= a_3 * b_3
\end{aligned}$$

The result, $c(x)$, is now reduced modulo a polynomial of degree 4. The calculations are given in Appendix A.1. In the Rijndael specification this polynomial is $M(x) = x^4 + 1$. The result is given by,

$$d(x) = d_3 x^3 + d_2 x^2 + d_1 x + d_0 \tag{1.13}$$

with,
$$\begin{aligned}
d_0 &= c_0 \oplus c_4 = a_0 * b_0 \oplus a_3 * b_1 \oplus a_2 * b_2 \oplus a_1 * b_3 \\
d_1 &= c_1 \oplus c_5 = a_1 * b_0 \oplus a_0 * b_1 \oplus a_3 * b_2 \oplus a_2 * b_3 \\
d_2 &= c_2 \oplus c_6 = a_2 * b_0 \oplus a_1 * b_1 \oplus a_0 * b_2 \oplus a_3 * b_3 \\
d_3 &= c_3 = a_3 * b_0 \oplus a_2 * b_1 \oplus a_1 * b_2 \oplus a_0 * b_3
\end{aligned}$$

This polynomial multiplication may be represented as,

$$\begin{bmatrix} d_0 \\ d_1 \\ d_2 \\ d_3 \end{bmatrix} = \begin{bmatrix} a_0 & a_3 & a_2 & a_1 \\ a_1 & a_0 & a_3 & a_2 \\ a_2 & a_1 & a_0 & a_3 \\ a_3 & a_2 & a_1 & a_0 \end{bmatrix} \begin{bmatrix} b_0 \\ b_1 \\ b_2 \\ b_3 \end{bmatrix} \tag{1.14}$$

1.6.5. Multiplicative Inverse in GF(2^8)

The multiplicative inverse of a polynomial in the finite field, GF(2^8) is found by performing the Extended Euclidean Algorithm. An irreducible polynomial of degree 8, which is defined as, $m(x) = x^8 + x^4 + x^3 + x + 1$ in the Rijndael algorithm, is required in the calculation.

Taking a polynomial $a(x)$, the Extended Euclidean Algorithm is performed on $a(x)$ and $f(x)$. If $a(x)$ is not zero, the polynomials $r(x)$ and $s(x)$ will be obtained such that,

$$r(x)*a(x) + s(x)*f(x) = 1 \tag{1.15}$$

If this result is reduced modulo $f(x)$,

$$r(x)*a(x) = 1(\mod(f(x))) \tag{1.16}$$

and $r(x)$ will be the multiplicative inverse of $a(x)$.

For example, the inverse of the byte 1100 1011 (0xCB) is found as described in Table 1-12.

The *Auxiliary* column commences with the values 0 and 1 in rows 1 and 2. Similarly, the *Remainder* column starts with $m(x)$ and $a(x)$.

Table 1-12. Multiplicative Inverse of 0xCB in $GF(2^8)$

Row	Remainder	Quotient	Auxiliary
1	$m(x) = x^8 + x^4 + x^3 + x + 1$	-	0
2	$a(x) = x^7 + x^6 + x^3 + x + 1$	-	1
3	$x^6 + x^2 + x$	$x + 1$	$x + 1$
4	1	$x + 1$	x^2

To fill the subsequent rows, divide the remainder, $m(x)$ by $a(x)$ and place the quotient in the *Quotient* column of row 3 and the remainder in the *Remainder* column of row 3. Next multiply the quotient in row 3 by the auxiliary value in the previous row, row 2 and add the result to the auxiliary value that appears in the row before that, row 1. Continue this process until the remainder is reduced to 1. The contents in the corresponding auxiliary column is the inverse of $a(x)$ (see Appendix A.2 for the full calculation). Hence, the inverse of $a(x) = x^7 + x^6 + x^3 + x + 1$ is equal to x^2.

1.7. Conclusions

The art of secret writing has been in existence since ancient times and many wars and battles have been won, not by the soldiers on the battlefield, but through the breaking of secret enemy codes and ciphers by mathematicians and scientists. Perhaps the most famous military use of encryption was during WWII, when the German Enigma codes were broken by scientists such as Alan Turing at Bletchley Park, providing the allies with vital information regarding their war efforts.

In this chapter both classical and modern cryptographic techniques were discussed. Classical encryption techniques typically involved character-based substitution and transposition ciphers, in which the letters of a message were rearranged and reorganised. More modern encryption systems, such as public and private key algorithms, hash functions and digital signatures, employ complex mathematics. Public key algorithms provide secure key distribution. The most commonly used public key cipher is RSA, which is based on the computationally infeasible problem of factoring large prime numbers. Private key algorithms are typically used for bulk data encryption. The best-known symmetric key cipher is the DES algorithm.

In 1997, DES was broken and in response, the NIST established the AES development effort to find a suitable algorithm to replace DES as the FIPS encryption standard. By August 1999, the NIST had chosen five finalists: MARS, Serpent, RC6, Rijndael and Twofish. In October 2000, after a long and rigorous evaluation process, Rijndael was chosen as the AES winner. Although each of the finalists appeared to offer adequate security, Rijndael was selected as it performed well over a wide range of hardware and software environments. It was one of the easiest algorithms to defend against timing and power attacks. It also had low memory requirements, making it highly suitable for memory-constrained applications. The Rijndael algorithm comprises substitutions, permutations and complex finite field mathematics. The finite field addition, multiplication and multiplicative inverse operations utilised in Rijndael are outlined in the chapter.

Passive and active cryptographic attacks have also been explained. There is a constant battle between cryptographers and cryptanalysts, without which, advances in the field of cryptography would never be achieved.

The advantages of hardware-based cryptographic architectures and in particular, FPGA implementations of encryption algorithms have been described. Hardware implementations significantly out-perform equivalent software implementations in terms of speed. They also support pipelined architectures and provide additional security since it is physically more difficult to tamper with hardware encryption devices. FPGAs offer many of the advantages of custom-built ASIC implementations such as higher throughput and security. Although ASICs provide a faster implementation and are more efficient to use in high-volume applications, FPGAs have a much shorter development time, lower NRE costs and support in-circuit reprogrammability [19]. Since they share many of the advantages provided by ASICs without the disadvantages, FPGA implementations are becoming a practical and viable alternative solution.

Chapter 2

DES ALGORITHM ARCHITECTURES AND IMPLEMENTATIONS

2.1. Introduction

The DES algorithm [35] is the best known and most widely used encryption algorithm. IBM developed DES in the mid 1970s as a modification of an earlier system known as Lucifer. In 1976, DES was adopted by the National Bureau of Standards (NBS), now the NIST, as a federal standard and authorised for use on all classified government communications. It has been the standard algorithm for banking and other applications since 1977. Although the private-key algorithm has recently been replaced by the Advanced Encryption Standard (AES) algorithm, DES will still remain in the public domain for a number of years because of legacy requirements. It provides a basis of comparison for new algorithms and it is also used in IPSec protocols, ATM cell encryption, the Secure Socket Layer (SSL) protocol and in Triple-DES, adopted to improve DES in the X9.17 and ISO 8732 standards [36, 37].

Much design work has been carried out on the DES algorithm. This design work has concentrated on high throughput designs and small, low area designs. This chapter describes a new generic parameterisable DES IP architecture. The use of this approach allows designs to be quickly created for a wide range of specifications in terms of scalability, modes of operation and single or Triple-DES functionality. A novel key scheduling design, which can be utilised in any pipelinable private-key encryption algorithm implementation, is also presented [38, 39, 40]. The design supports the use of different keys every clock cycle, thus improving overall security since users are not restricted to using the same key during any one session of data transfer. The DES algorithm, which lends itself readily to pipelining, is

DES Algorithm Architectures and Implementations 29

utilised to exemplify this novel scheduling method. The broader applicability of the method to other encryption algorithms is also illustrated. Both designs are implemented on Xilinx Virtex FPGA technology [24].

The chapter begins with a detailed description of the DES algorithm. The Triple-DES algorithm and DES modes of operation are also discussed. Previous design work on the DES algorithm is then outlined. The generic, parameterisable DES design is described and performance results evaluated. The design and a performance analysis of the novel key scheduling core is also examined. Finally, in the conclusion, a comparison is presented between previous work and the designs discussed in the chapter.

2.2. DES Algorithm Description

DES is a private key (symmetric) algorithm. An outline of DES is shown in Figure 2-1. It is a block cipher operating on 64-bit blocks of plaintext utilising a 64-bit key. Every 8th bit of the 64-bit key is used for parity checking and otherwise ignored. After an initial permutation, the 64-bit input is split into a right half, R_o, and a left half, L_o, each 32 bits in length. DES has 16 iterations or rounds. In each round a function, f, is performed in which the data is combined with a 48-bit permutation of the key. After the 16th iteration, the right (R_{16}) and left (L_{16}) halves are concatenated and a final permutation, which is the inverse of the initial permutation, completes the algorithm.

2.2.1. Function f of the DES Algorithm

The function f of the DES algorithm is made up of four operations. Firstly, the 32-bit right half of the plaintext, R_o, is expanded to 48-bits and then XORed with a 48-bit sub-key, K_1. The result is fed into eight substitution boxes (S-Boxes), which transform the 48-bit input to a 32-bit output. Finally, a straight permutation (P-Permutation) is performed, the output of which is XORed with the initial left half, L_o, to obtain the new right half, R_1. The original right half, R_o, becomes the new left half, L_1. This is outlined in Figure 2-2.

Expansion and P-Permutation

In the expansion permutation, the right half of the input data is expanded from 32 to 48 bits. This is achieved by repeating sixteen of the bits and rearranging their order. For example, bit 1 of the input will form bit 2 and the final bit of the 48-bit output. The P-Permutation simply involves rearranging the order of its 32-bit input to achieve a transformed 32-bit output. Both permutations are outlined in Appendix B.1.

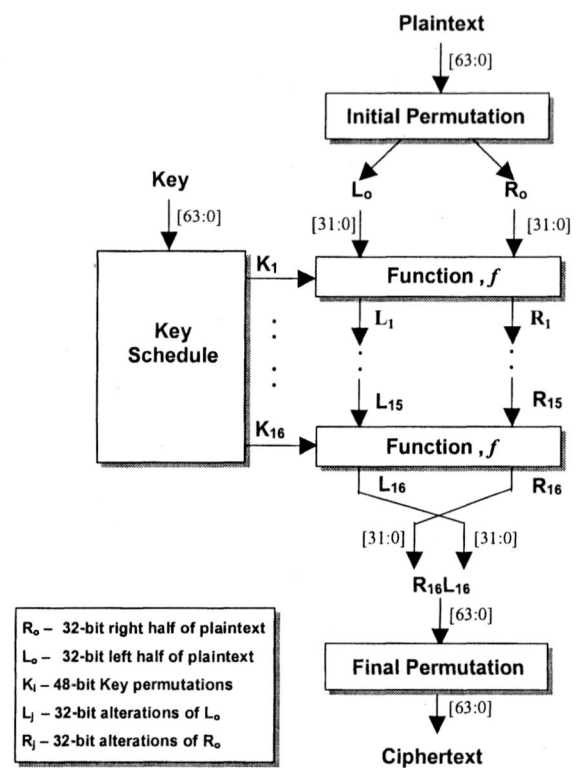

Figure 2-1. Outline of DES Encryption Algorithm

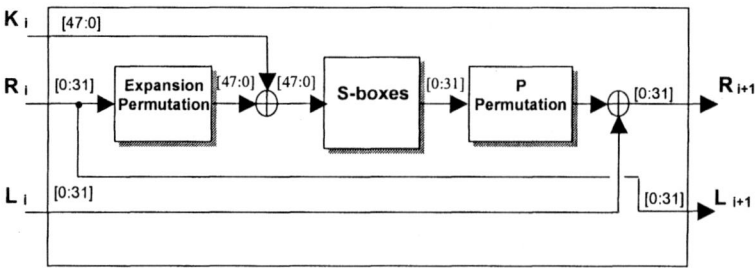

Figure 2-2. Function, f of the DES Algorithm

Substitution Boxes (S-Boxes)

Each S-Box has a 6-bit input and a 4-bit output. Hence the 48 bits are divided into 6-bit sub-blocks and each of these is operated on by an S-Box. An S-Box consists of a table of 4 rows and 6 columns, where the entries are integers in the range 0 to 15 as specified in the standard. The 6-bit input specifies in which row and column the 4-bit output value can be obtained. The outputs from each of the 8 S-Boxes are concatenated to obtain a 32-bit output.

Consider an input: $x_1\ x_2\ x_3\ x_4\ x_5\ x_6$
- x_1 and x_6 combine to give a 2-bit number in the range 0 to 3, which identifies the appropriate row.
- $x_2\ x_3\ x_4\ x_5$ combine to form a 4-bit number in the range 0 to 15, which identifies the appropriate column.

For example, if the input to S-Box 6, outlined in Table 2-1, is '*1*100*1*', then combining x_1 and x_6 gives '11', i.e. 3; combining $x_2\ x_3\ x_4\ x_5$ gives '1001', i.e. 9. Therefore the required output is located in row 3, column 9. From the table this entry is 14 or '1110' when represented in binary, and thus the 6-bit input is reduced to a 4-bit output. All eight S-Boxes are given in Appendix B.2.

Table 2-1. DES S-Box 6

	0	1	2	3	4	5	6	7	8	9	10	11	12	13	14	15
0	12	1	10	15	9	2	6	8	0	13	3	4	14	7	5	11
1	10	15	4	2	7	12	9	5	6	1	13	14	0	11	3	8
2	9	14	15	5	2	8	12	3	7	0	4	10	1	13	11	6
3	4	3	2	12	9	5	15	10	11	*14*	1	7	6	0	8	13

2.2.2. Key Scheduling

There are two methods of implementing the DES key procedure. The initial step in the first method, the *shifting* method, is to remove the parity check bits in the 64-bit key. Every 8th bit is used for parity checking, leaving 56-bits. The permutation used to remove the parity bits is provided in Appendix B.3. A different 48-bit sub-key is now generated for each of the 16 rounds of DES. The sub-keys are determined by first splitting the 56-bits into two 28-bit lengths of data. Then both halves are shifted left by either one or two bits depending on the round number, as outlined in Table 2-2. Finally, 48 of the 56 bits are selected according to the compression permutation shown in Table 2-3. The sub-key used in each iteration will be different due to the shifting operation. The second method, the *permutation* method, simply involves the implementation of the resulting 48-bit permutations.

Table 2-2. Number of Bits in Sub-Key to be Shifted each Round

Round	1	2	3	4	5	6	7	8	9	10	11	12	13	14	15	16
Shift	1	1	2	2	2	2	2	2	1	2	2	2	2	2	2	1

Table 2-3. Key Compression Permutation

14	17	11	24	1	5	3	28	15	6	21	10	23	19	12	4
26	8	16	7	27	20	13	2	41	52	31	37	47	55	30	40
51	45	33	48	44	49	39	56	34	53	46	42	50	36	29	32

2.2.3. DES Decryption

DES decryption employs the same algorithm as in the encryption process, but the 48-bit sub-keys created are used in reverse order. The first round utilises the final key, the second round uses key 15 and so on. The symmetry of the algorithm is evident in Figure 2-3, which illustrates the encryption and decryption processes, since,

$$L_0 = R_1 \oplus f(L_1, K_1) \equiv R_1 = L_0 \oplus f(R_0, K_1) \qquad (2.1)$$

Also, the final permutation is an inverse of the initial permutation. Therefore, in decryption its effect is cancelled by the initial permutation.

2.3. DES Modes of Operation

The Federal Information Processing Standards Publications (FIPS PUBS) 81 [41] defines modes of operation for the DES algorithm which may be used in a wide variety of applications. The four standard modes are the Electronic Codebook (ECB) mode, the Cipher Block Chaining (CBC) mode, the Cipher Feedback (CFB) mode and the Output Feedback (OFB) mode.

DES Algorithm Architectures and Implementations

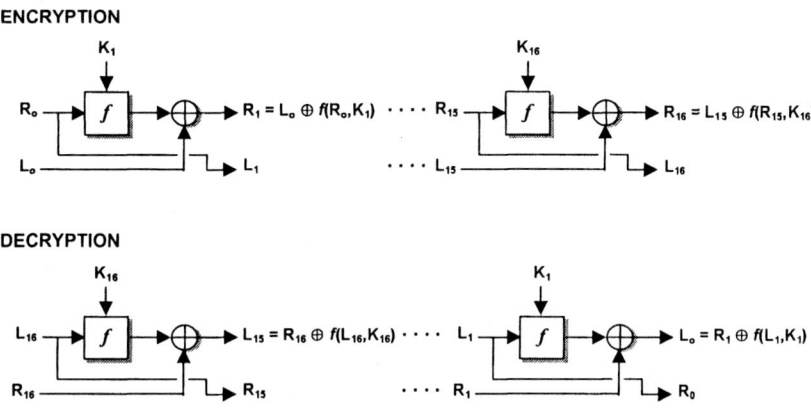

Figure 2-3. DES Encryption and Decryption Processes

2.3.1. Electronic Codebook (ECB) Mode

The ECB mode is the mode in which the DES algorithm is most commonly utilised. It involves 64-bit blocks of data being encrypted by the DES algorithm to obtain 64-bit blocks of ciphertext. The ciphertext blocks are then sent to the recipient of the message where they are decrypted to give the original data. The ECB mode is illustrated in Figure 2-4. A disadvantage of DES in ECB mode is that the security of the system relies entirely on the secrecy of the key. In order to achieve better security, DES can be used in one of the other three modes.

Figure 2-4. ECB Mode

2.3.2. Cipher Block Chaining (CBC) Mode

In CBC mode, an initial 64-bit block of data is used, known as an *initialisation vector* (*IV*). Initially, the *IV* is XORed with the first block of plaintext, P_0, and the result is encrypted to obtain the ciphertext block, C_0. Subsequently, ciphertext blocks, C_{i-1}, are XORed with plaintext blocks, P_i, prior to encryption. When decrypting the process is reversed, as shown in Figure 2-5. The initial vector can be regarded as an extra key, which guarantees extra security.

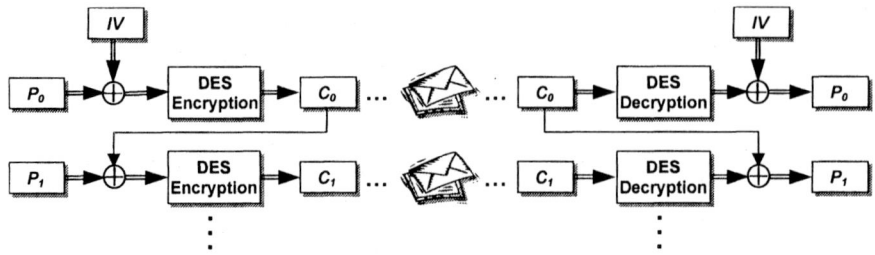

Figure 2-5. CBC Mode

2.3.3. Ciphertext Feedback (CFB) Mode

In CFB mode DES operates as a stream cipher. This mode also requires an initialisation vector. The *IV* is encrypted utilising the DES algorithm and *n*-bits of the result are XORed with *n*-bits of the plaintext to produce *n*-bits of ciphertext. The *n*-bit ciphertext is fed back to form part of the input to be used in the next encryption as shown in Figure 2-6. After the first encryption the initial input will be shifted left by *n*-bits and the next input therefore will consist of a *(64 − n)*-bit data block concatenated with the *n*-bits of the ciphertext. Decryption in CFB mode also uses DES encryption and *n*-bits of the result are XORed with *n*-bits of the ciphertext to produce *n*-bits of the original plaintext.

2.3.4. Output Feedback (OFB) Mode

OFB mode also allows DES to operate as a synchronous stream cipher. This mode is similar in structure to the CFB mode as depicted in Figure 2-7. The *IV* is again encrypted utilising the DES algorithm and *n*-bits of the result are XORed with *n*-bits of the plaintext to produce *n*-bits of ciphertext. However, *n*-bits of the result rather than the ciphertext are fed back to form part of the input to be used in the next encryption. Similarly to CFB mode, decryption in OFB mode uses DES encryption and *n*-bits of the result are XORed with *n*-bits of the ciphertext to produce *n*-bits of the original plaintext.

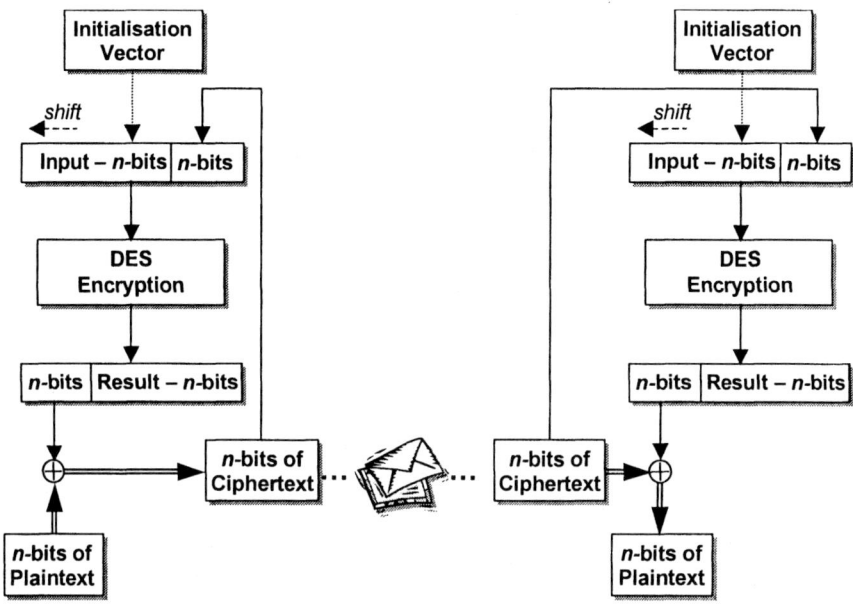

Figure 2-6. CFB Mode

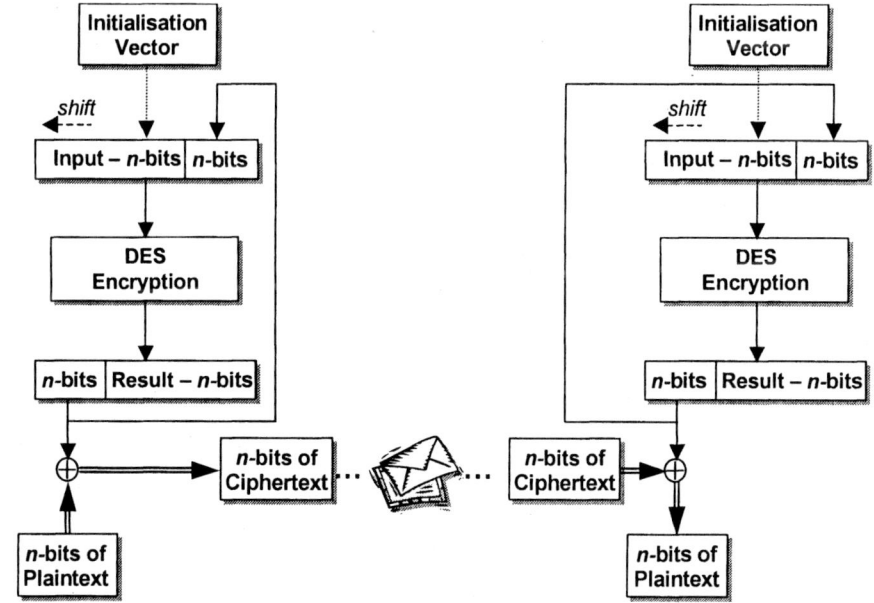

Figure 2-7. OFB Mode

2.3.5. Mode Advantages and Disadvantages

Advantages, disadvantages and a typical application [9] for the ECB, OFB, CFB and CBC modes of operation are as follows:

ECB Mode
Advantages:
- More than one message can be encrypted with the same key
- Processing is parallelisable

Disadvantages:
- Plaintext patterns are not concealed
- A ciphertext error affects one full block of plaintext

Typical Application:
- Secure transmission of small amounts of data such as an encryption key

CBC Mode
Advantages:
- Plaintext patterns are concealed by XORing with previous ciphertext block
- More than one message can be encrypted with the same key

Disadvantages:
- A ciphertext error affects one full block of plaintext and the corresponding bit in corresponding blocks

Typical Application:
- General-purpose block oriented transmission
- Authentication

CFB Mode
Advantages:
- Plaintext patterns are concealed
- More than one message can be encrypted with the same key, provided a different initialisation vector is used

Disadvantages:
- A ciphertext error affects the corresponding bit of plaintext and the next full block

Typical Application:
- Stream oriented transmission
- Authentication

OFB Mode
Advantages:
- Plaintext patterns are concealed
- More than one message can be encrypted with the same key, provided a different initialisation vector is used
- A ciphertext error affects only the corresponding bit of plaintext

Disadvantages:
- Vulnerable to message stream modification attacks since controlled changes can be made to recovered plaintext

Typical Application:
- Stream oriented transmission over noisy channels such as satellite communication

2.4. Triple-DES

The Triple Data Encryption Algorithm (TDEA) [35] was approved by the NIST as the symmetric encryption algorithm of choice over DES in October 1999. The deployment of TDEA, more commonly referred to as Triple-DES, took place in response to the security problems encountered by DES.

Triple-DES uses three DES encryption/decryption operations on data, with three different keys. Encrypting data with Triple-DES involves performing DES encryption using the first key, K_1, the result is then decrypted with the second key, K_2 and finally, this result is encrypted using the third key, K_3. Triple-DES encryption and decryption are illustrated in Figure 2-8.

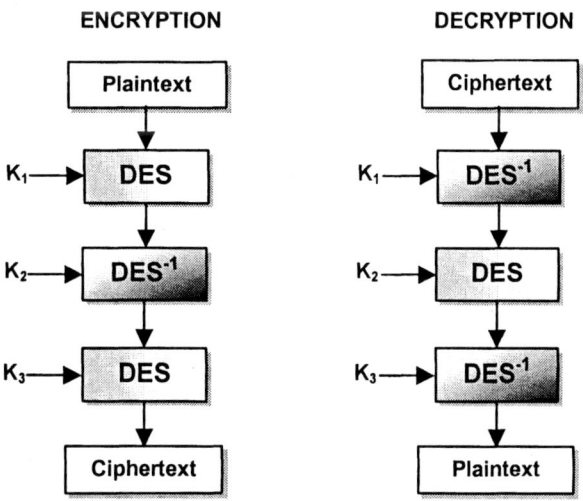

Figure 2-8. Triple-DES Encryption and Decryption

The TDEA algorithm standard defines three keying options for the three keys, K_1, K_2 and K_3:
- K_1, K_2 and K_3 as independent keys (this is in effect a 168-bit key)
- K_1 and K_2 as independent keys and $K_3 = K_1$
- All three keys are the same, $K_1 = K_2 = K_3$

The most common implementation of Triple-DES employs the second keying option and is known as *encrypt-decrypt-encrypt* or *EDE2* mode. The DES modes of operation, ECB, CBC, OFB and CFB, also apply to Triple-DES [42] and simply involve substituting the DES algorithm with the Triple-DES algorithm.

2.5. Review of Previous Work

ASICs offer the highest performance DES hardware implementations but lack flexibility. The design by Goubert *et al.* [43] is one of the earliest references to a custom hardware implementation of DES. The maximum speed of the chip was 20 Mbits/sec. A later implementation on a gallium arsenide (GaAs) gate array [44] achieved an encryption rate of 1 Gbit/sec. The Sandia National Laboratories (SNL) DES implementation is the fastest known ASIC implementation, capable of running at 9.28 Gbits/sec. The ASIC was fabricated using 0.6 micron CMOS technology [45].

Leonard and Mangione-Smith [46] published one of the first papers on FPGA implementations of the DES algorithm in 1997. Their fastest implementation achieves a data-rate of 26.4 Mbits/sec on a Xilinx XC4000 series device. However, the design does not support decryption and each key must be pre-computed before it can be used in the device. A single-chip implementation of DES on a Xilinx XC4000 series device is described by Wong *et al.* [47]. Their design achieves an encryption speed of 26.7 Mbits/sec. Kaps and Paar [48] carried out extensive research on high-speed FPGA architectures for the DES algorithm. In their studies they consider a pipelined design with a four-stage pipeline. The data-rate achieved is 402.7 Mbits/sec on the XC4028EX device. They also achieved a 99 Mbits/sec iterative design on a XCV4008E device. A JBits implementation of DES by Patterson [49] performs at 10.7 Gbits/sec on a Virtex XCV150 device. JBits provides a Java-based Application Programming Interface (API) for the run time creation and modification of the configuration bitstream, which allows dynamic circuit specialisation based on a specific key and mode. In this implementation, the key schedule is computed entirely in software and forms part of the bitstream. Therefore, all key input and subkey generation circuitry is removed. This design is not a single-chip implementation of the full DES algorithm since the key schedule is computed in software. Also, it can only accommodate one key per data transfer session. A free-DES core [50] also

DES Algorithm Architectures and Implementations

exists which utilises a 16-stage pipelined DES design implemented on a Virtex XCV400-6 device. It achieves a throughput of 3.05 Gbits/sec. The fastest known DES FPGA implementation is by Trimberger, Pang and Singh [51] and runs at 12 Gbits/sec on an XCV300E FPGA. This design is heavily pipelined and optimised specifically for the device on which it is implemented.

Although software implementations can easily convert between encryption algorithms, they suffer poor data-rates. The highest performance of a software implementation is by Eli Biham [52]. The implementation achieves a throughput of 137 Mbits/sec on a 300 MHz Alpha 8400 processor.

Table 2-4. Specifications for Recent DES Implementations

Manufacturer	Type of Design	Device Used	Area	System Clock (MHz)	Data Rate Mbits/sec
Kaps, Paar [48]	Iterative	XC4008E	524 slices	40.4	99
Kaps, Paar [48]	Pipelined (x4)	XC4028EX	1482 slices	39.7	403
Biham [52]	-	Alpha 8400	-	300	137
Sandia Laboratories [45]	Pipelined(x16)	CMOS	256,000 transistors	145	9280
Patterson [49]	Pipelined(x16)	XCV150	1584 slices	168	10752
Trimberger et al. [51]	Pipelined(x16)	XCV300E	72952 gates	187.5	12000

Table 2-4 summarises the performance of the fastest implementations outlined above and for comparison purposes the specifications for a number of commercial hardware implementations of the DES algorithm are given in Table 2-5.

Table 2-5. Specifications for Commercial DES Hardware Implementation

Manufacturer	Type of Design	Device Used	CLB Slices	System Clock (MHz)	Data Rate Mbits/sec
Memec [53]	Iterative	XCV4013XL	316	43	172
CAST Inc [54]	Iterative	XCV150-6	255	101	404
Free-DES [50]	Iterative	XCV400	905	32.5	130
Free-DES [50]	Pipelined (x16)	XCV400	5263	47.7	3052

40   *Chapter 2*

2.6. Generic Parameterisable DES IP Architecture Design

The DES algorithm is over 20 years old and therefore many designs exist which have been forged to specific application requirements. The aim of the research presented in this section is to describe a new generic architecture and design, which accommodates all application requirements.

The architecture uses generic parameters to allow the following:
- ECB or CBC modes of operation
- Technology-independent, Virtex FPGA or Virtex-II FPGA implementation of the DES S-Boxes
- Single or Triple-DES functionality
- Two methods of key generation
- A varying number of pipeline stages - 1, 2, 4, 8 or 16

A finite state machine is used to control the overall functionality of the design. Figure 2-9 illustrates the VHSIC Hardware Description Language (VHDL) package file containing the generic parameters.

```
library IEEE;
use IEEE.std_logic_1164.all;

    package DES_PACK is

        -- METHOD OF GENERATING KEYS
        -- 1 indicates permutation method, 0 indicates shifting method
        constant KEY_TYPE : integer := 0;


        -- ENABLE USE OF DISTRIBUTED RAM PRIMITIVES
        -- 1 enables use of distributed RAM for the s-boxes, 0 uses case statement
        constant USE_SBOX_DRAM : integer := 1;


        -- NUMBER OF ITERATIONS TO BE IMPLEMENTED
        -- indicates number of rounds implemented (1,2,4,8 or 16 are all valid)
        constant NUM_ITERATE : integer := 16;


        -- TARGET VIRTEX TECHNOLOGY
        -- 0 for virtex , 1 for virtex II implementation
        constant VIRTEX_TYPE : integer := 1;


        -- ENABLE TRIPLE DES IMPLEMENTATION
        -- 0 for DES, 1 for Triple DES (EDE2 mode)
        constant TRIPLE_DES : integer := 0;


        -- DES MODE IF ONE ITERATION TO BE IMPLEMENTED
        -- 0 for ECB, 1 for CBC
        constant MODE : integer := 0;

    end DES_PACK;
```

Figure 2-9. VHDL Package file Containing Generic Parameters

DES Algorithm Architectures and Implementations 41

2.6.1. Pipelining the DES Core

The iterative nature of the DES algorithm makes it ideally suited to pipelining. DES comprises 16 rounds. Therefore, a high-speed hardware DES design can be achieved by implementing a 16-stage fully pipelined architecture. However, such a design incurs a large area penalty. If low silicon area is a requirement, it is possible to feed data through just one DES round in an iterative process, but this design will lead to low data throughputs. In order to accommodate various area and speed requirements, the DES architecture described in this section is designed to support 1, 2, 4, 8 or 16 pipeline stages.

In order to accomplish parameterisation of the number of pipeline stages to be implemented, an initial architecture containing the full 16 stages is designed. Parameters are then used to select the number of rounds in the architecture required for synthesis. This design can be coded using the VHDL *generate* statement and generic parameters. Separate state machines were designed for the 2, 4 and 8-stage pipelined designs and a *generate* statement was utilised to instantiate the correct state machine. Figure 2-10 outlines the structure of DES with two pipeline stages. Registers are placed at the left and right outputs of each function f block to allow the data to be sequenced.

2.6.2. Permutations

The initial, final, expansion and straight permutations of the DES algorithm involve a rearrangement of their input data and thus, are simply hardwired as no logic is involved in their implementation.

2.6.3. Key Generation

Both key generation methods have been included in the architecture. The design of the *permutation* method involves the rearrangement of the input key data, similar to the other DES algorithm permutations. However, when utilising this method only one key can be used in any one data transfer session.

The *shifting* method of implementing the DES key schedule involves carrying out a shift operation during each algorithm iteration. Figure 2-11 outlines the shifting key scheduling method for a fully pipelined architecture.

The 64-bit input key is passed through an initial key permutation, which removes the parity check bits. The resulting 56-bits enters the first function f block. It is split into two 28-bit words and a cyclic shift left operation is carried out on each half. The two halves are then concatenated and operated on by a final key permutation to obtain the sub-key required by the function f block. The output from the shift operation provides the input to the next function block where once again a cyclic shift left operation is carried out and a key permutation performed. This process continues for each iteration. In rounds 1, 2, 9 and 16 of the DES algorithm the halves are shifted one position to the left and for all other rounds two positions to the left.

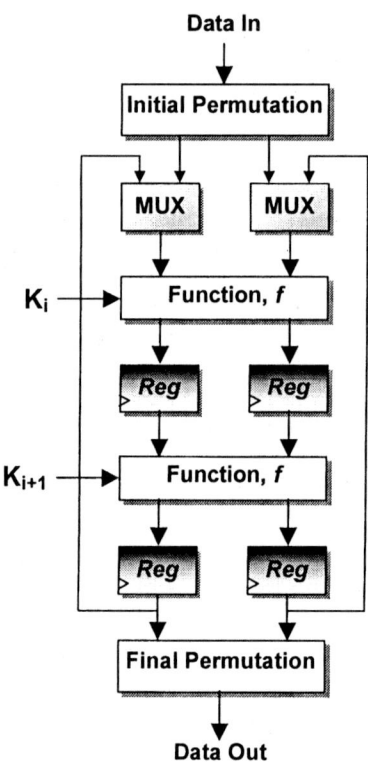

Figure 2-10. DES with Two Pipeline Stages

DES Algorithm Architectures and Implementations

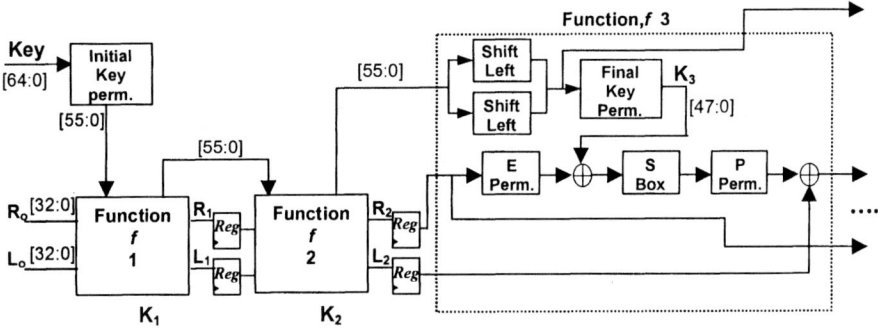

Figure 2-11. Shifting Key Scheduling Method for a Pipelined Architecture

The shifting method when utilised with pipelined designs should use less on-chip routing resources than the permutation method, and thus should improve overall throughput. A disadvantage of this method, however, is the initial latency incurred in the generation of the sub-keys. During encryption the latency is decreased as final sub-keys need only be generated and available when the data reaches the final rounds. For decryption the latency is 16 clock cycles, since the sub-keys are utilised in reverse order and thus, must be pre-generated. However, the permutation method should produce a smaller design since no actual logic is involved in its implementation.

2.6.4. Design of DES S-Boxes

Three different options are available for implementation of the DES S-Boxes – technology-independent implementation, specific implementation for a Virtex FPGA device or specific implementation for a Virtex-II FPGA. Each S-Box is, in effect, a LUT with a 6-bit input and 4-bit output. Therefore, for the technology-independent design, these look-up tables are simply coded using *case* statements.

Distributed RAM was utilised in the specific implementation of the S-Box LUTs in the Virtex and Virtex-II devices. A study by Haskins [55] indicates that using ROM blocks provides the most efficient implementation for the S-Boxes of the DES algorithm. Distributed RAM is located within each CLB slice on a Virtex or Virtex-II FPGA. Hence, the S-Boxes will be in close proximity to the remainder of the DES algorithm's implemented components, thus reducing signal propagation delays. Propagation delays arise when data has to travel long distances across a chip to dedicated memory resources, such as the Block RAM. When the write enable of the distributed RAM is low ('0'), transitions on the write clock are ignored and data stored in the RAM is not affected. Hence, if the RAM is initialised and both the input data and write enable pins are held low then the RAM can be utilised as a ROM or LUT.

The maximum size of distributed RAM available on a Virtex device is a 32 x 1-bit RAM. Each S-Box comprises 256 bits of data and therefore eight 32 x 1-bit RAM are required in an implementation. Figure 2-12 outlines the structure of an S-Box when implemented on a Virtex device. 5 bits of the 6-bit input forms each RAM address (A4 to A0). The final 1 bit is utilised as the select signal for each multiplexor.

One advantage of using a Virtex-II over a Virtex FPGA in the implementation of DES S-Boxes is that in Virtex-II devices, 64 x 1-bit RAM are available. Therefore, only four 64 x 1-bit RAM are required in the implementation of each S-Box, as depicted in Figure 2-13. This will lead to a more compact design in comparison to a Virtex implementation. For example, a 16-round DES design will utilise 256 CLBs (2 CLBs per S-Box) to implement the S-Boxes on a Virtex-II device compared to 512 (4 CLBs per S-Box) on a Virtex FPGA.

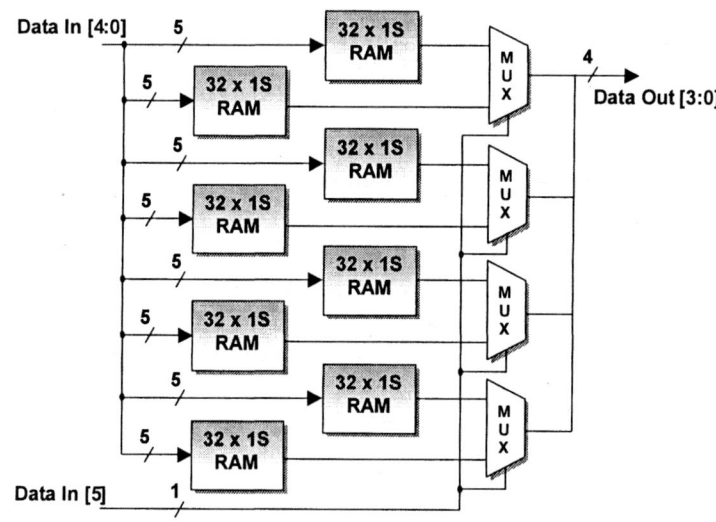

Figure 2-12. Structure of S-Box Implemented on a Virtex Device

DES Algorithm Architectures and Implementations 45

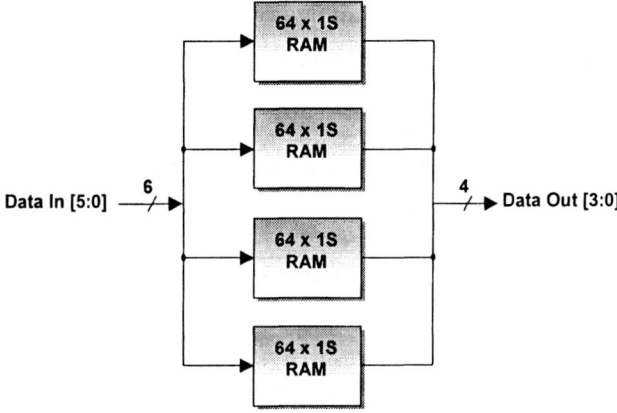

Figure 2-13. Structure of S-Box Implementation on a Virtex-II Device

2.6.5. Triple-DES and DES Mode Options

The generic DES architecture is designed so that the algorithm can be performed three times in succession, and thus supports the implementation of Triple-DES. If the generic parameter, TRIPLE-DES is assigned high, three DES entities are implemented with logic to support a 116-bit key in accordance with the *EDE2* mode. The DES architecture supports a non-feedback (ECB) and a feedback (CBC) mode of operation and can easily be modified to support other feedback modes. Since CBC mode is a feedback mode, it can only be performed with the 1-stage pipeline design. The ECB is a non-feedback mode and thus its operation can be pipelined [13]. A fully pipelined DES implementation will also operate in counter mode. Counter mode is a simplification of Output Feedback (OFB) mode and involves updating the input plaintext block as a counter, $I_{j+1} = I_j + 1$, rather than using feedback. Hence, the ciphertext block, i is not required in order to encrypt plaintext block, $i+1$ [29]. Counter mode provides more security than ECB mode, however, operation of either mode involves trading security for high throughput.

2.6.6. Performance Evaluation

In order to evaluate the performance of the generic DES architecture, two characteristics are considered – the area utilised and the throughput achieved by the designs. Firstly, the performance of the 1, 2, 4, 8 and 16-stage pipelined designs were investigated. Each of the designs was implemented using the shifting key scheduling method, on XCV400-4 Virtex FGPA devices. The performance results obtained are outlined in Table 2-6.

Table 2-6. Performance Results for Pipelined DES Cores Implemented on Virtex XCV400 Devices

Number of Pipeline Stages	Throughput (Mbits/sec)	Number of CLB Slices	Throughput / CLB Slice ($\times 10^3$)	% Relative Area Efficiency	% Relative Speed Efficiency
1	213.4	336	635	100	5
2	394.5	1215	324.7	27.7	9.3
4	615.6	1967	312.9	17	14.6
8	904.5	3066	395	10.9	21.4
16	4227	2844	1486	11.8	100

The 8-stage pipelined design is in fact larger in area than the fully pipelined design. This is because the feedback logic which is required to control the data flow and the key usage in the 8-stage design is more area inefficient than the implementation of the eight additional iterations.

To gain an appreciation of the area versus throughput trade-off, the relative performance of the pipelined designs are plotted in Figure 2-14, where η_A is defined as the relative area efficiency and η_T is the relative throughput efficiency such that,

$$\eta_A = \text{minimum area / actual area} \qquad (2.2)$$

and,

$$\eta_T = \text{actual throughput / maximum throughput} \qquad (2.3)$$

The 16-stage fully pipelined architecture is the most efficient of the pipelined designs. Interestingly, the iterative design is more efficient than the 2, 4 and 8-stage pipelined designs since it is a much more compact core and yet achieves a desirable throughput. Indeed, using Figure 2-14, the generic DES architecture can meet the specification of any application. If the area and throughput requirements are weighted, the most cost-effective design for that application can be determined.

To illustrate the effect of utilising the shifting and permutation key scheduling methodologies with pipelined architectures, the 4 and 16-stage pipelined designs were implemented using both techniques on Virtex XCV400 FPGAs. The throughput and area figures achieved are illustrated in Table 2-7.

DES Algorithm Architectures and Implementations

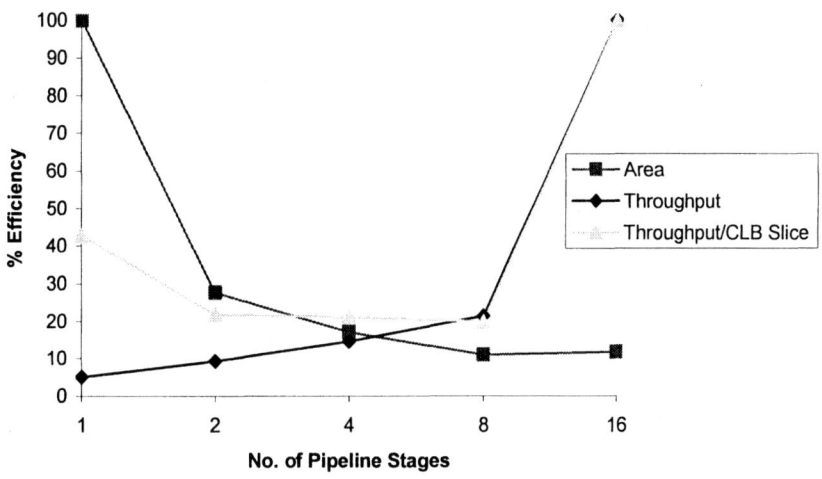

Figure 2-14. Area and Throughput Efficiency of Pipelined DES Cores

Table 2-7. Performance Results of Pipelined DES Cores Utilising Shifting and Permutation Key Scheduling Techniques

Number of Pipeline Stages	Key Scheduling Method	Throughput (Mbits/sec)	Number of CLB Slices	Throughput / CLB Slice $(\times 10^3)$ (Mbits/sec*Slices)
4	Shifting	615.6	1967	313
	Permutation	576	1885	305.5
16	Shifting	4227	2844	1486.3
	Permutation	3737	2714	1377

As expected, the shifting method yields higher throughputs than the permutation method at the expense of area. Overall, the shifting technique proves to be the more efficient of the two key scheduling systems. Therefore, if a high-throughput design is required which involves few key changes, the shifting key generation method should be utilised. However, if frequent key changes are necessary, the permutation method should be employed since no initial latency is incurred when a new key is loaded.

The Triple-DES design performs at a clock speed of 48 MHz when implemented on the XCV400 Virtex device. It operates at a data-rate of 64 Mbits/sec and utilises 1378 CLB slices. Thus, it is approximately three times the area of the iterative design.

Table 2-8. Performance Results of DES Cores Implemented on Virtex, Virtex-E and Virtex-II Devices

Number of Pipeline Stages	Device	Throughput (Mbits/sec)	Number of CLB Slices	Throughput / CLB Slice (x10^3) (Mbits/sec*Slices)
1	Virtex XCV400-4	213.4	336	635
	Virtex-II XC2V500-4	278.3	281	990
	Virtex-E XCV300E-8	273.3	336	813.4
16	Virtex XCV400-4	4227	2844	1486.3
	Virtex-II XC2V500-4	6314.8	2596	2432.5
	Virtex-E XCV300E-8	7810.6	2713	2879

Finally, the iterative and fully pipelined designs were also implemented on Virtex-II and Virtex-E FPGA devices. The implementation results obtained are given in Table 2-8. When the designs are implemented on the enhanced Virtex-E and Virtex-II FPGA technologies significant improvements in speed are achieved. The fully pipelined DES core implemented on the Virtex-E device runs at a clock speed of 122 MHz and achieves a data throughput of 7.8 Gbits/sec, which is faster than that obtained on the Virtex-II device. However, a –4 speed grade Virtex-II device was utilised in comparison with a –8 speed grade Virtex-E FPGA. Therefore, higher throughputs can be achieved if faster speed grade Virtex-II devices are used for implementation. The CLBs in Virtex-II FPGAs comprise 64 x 1-bit distributed RAM and large multiplexors, which aid in creating a more compact implementation. Thus the Virtex-II implementations produce smaller overall designs.

2.7. Novel Key Scheduling Method

Two typical DES key generation techniques were incorporated in the design outlined in §2.6. In this section a novel implementation of the DES algorithm key schedule is presented which applies specifically to a pipelined design. It is an extension to the permutation method, which allows the loading of a different key every clock cycle. The sub-keys are pre-computed

DES Algorithm Architectures and Implementations 49

and hence, for a 16-stage pipelined DES design, it is necessary to control the time at which the sub-keys are available to each function f block. This is accomplished by the addition of a skew that delays the individual sub-keys by the required amount. An outline of this key scheduling method is provided in Figure 2-15. The design comprises two components, a sub-key generation block and a skew core.

2.7.1. Sub-key Generation

The new sub-key generation block developed constructs the key permutations required for each iteration of the DES algorithm. Each sub-key permutation has the same 64-bit key input (the initial key input) but a different 48-bit key output and simply involves a rearrangement of the input data. For example, the permutation required to create the first subkey during encryption is outlined in Figure 2-16. From this table it can be seen that bit 2 of the 64-bit input key becomes bit 9 of the 48-bit permutated key, bit 62 becomes bit 46, bit 63 becomes bit 40, and so on.

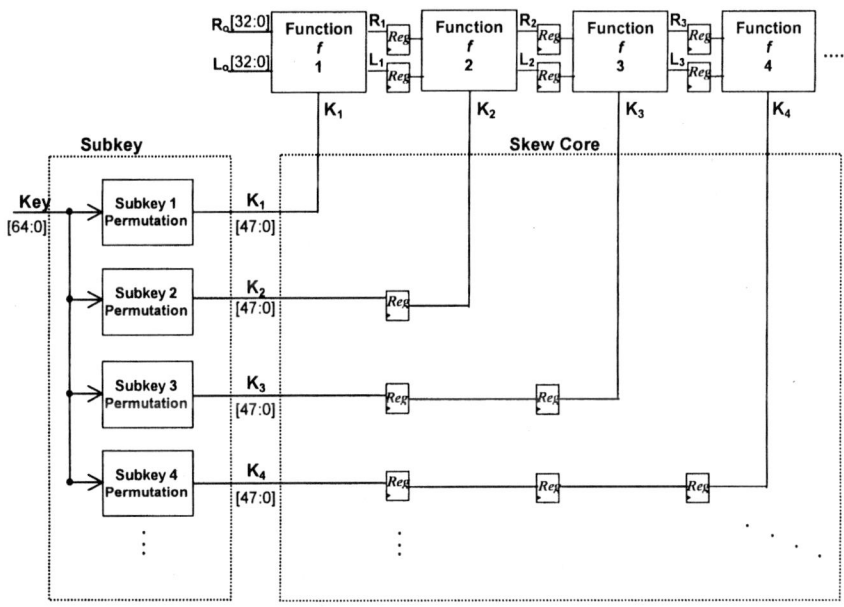

Figure 2-15. Outline of Novel Key Scheduling Method

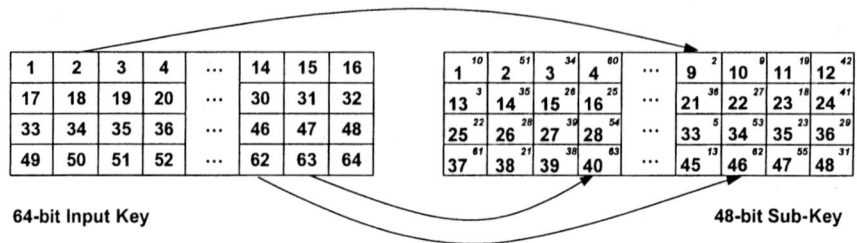

Figure 2-16. Round 1 Encryption Key Permutation

This key scheduling design also supports decryption. When decrypting data the keys for each round or iteration are used in reverse order. Therefore, sub-key permutation 1 will be used to create the first sub-key, K_1, in the encryption process and the final sub-key, K_{16}, in the decryption process.

2.7.2. Skew Core

The skew core consists of a 'dffarray' sub-component, which generates a sequence of registers as required. The code for this sub-component is outlined in Figure 2-17. The *Depth* parameter is generic and indicates the desired length of the array. If *Depth = 0*, the process, *S1* is used to create one register. If *Depth > 0*, for example if *Depth = 2* (effectively the Depth parameter will begin at 0 and count up to 2), the *S1* process is used to create the first register in the array and the *S2* process creates the remaining two registers. The 'skew' component generates an array of registers of varying lengths. It uses the 'dffarray' sub-component to produce the correct number of registers required at each round of the DES algorithm. The code for the 'skew' component is shown in Figure 2-18. Since the DES algorithm consists of 16 rounds, the skew core is set to loop 15 times *(for i = 0 to 14)* since a register is not required to delay the first sub-key. The value of *i* determines the *Depth* of the array to be generated by the 'dffarray' component. When *i = 0*, one register is created. If *i = 2* for example, three registers are created. Hence an array of registers of varying lengths is generated as illustrated in Figure 2-19.

DES Algorithm Architectures and Implementations 51

```
library IEEE;                                    signal A   : MIDKEY;
use IEEE.std_logic_1164.all;                     begin
use IEEE.std_logic_unsigned.all;                    G1: for i in 0 to Depth generate
use work.TYPES.all;                                    begin
                                                       S1: if i = 0 generate
--KEY is of type std_logic_vector(0 to 47)                FF1: dff1
entity dffarray is                                        port map(  D      => Keyin,
    generic ( Depth   : POSITIVE);                                   clk    => clk,
    port     ( Keyin  : in KEY;                                      reset  =>reset,
               clk    : in std_logic;                                Q      => A(I));
               reset  : in std_logic;                  end generate S1;
               D_Key  : out KEY);
end dffarray;                                          S2: if i /= 0 generate
                                                          FF2 : dff1
architecture synth of dffarray is                         port map(  D      => A(I-1),
                                                                     clk    => clk,
component dff1                                                       reset  => reset,
    port     ( D      : in std_logic_vector( 0 to 47);                Q      => A(I));
               clk    : in std_logic;                  end generate S2;
               reset  : in std_logic;
               Q      : out std_logic_vector( 0 to 47));  end generate G1;
end component;
                                                    D_Key <= A(Depth);
type MIDKEY is array (0 to Depth) of KEY;
                                                 end synth;
```

Figure 2-17. Code for 'Dffarray' Component

2.7.3. Applicability to Private-Key Algorithms

The novel key scheduling can be utilised in the implementation of any pipelinable private-key encryption algorithm. The design is particularly suited to substitution-permutation (SP) and feistel-structured algorithms. A feistel cipher utilises multiple iterations of a simple non-linear function as illustrated in Figure 2-20. DES is an example of a feistel cipher. As described in §1.5.3, an SP algorithm is one composed of a number of stages each involving substitutions and permutations.

52   *Chapter 2*

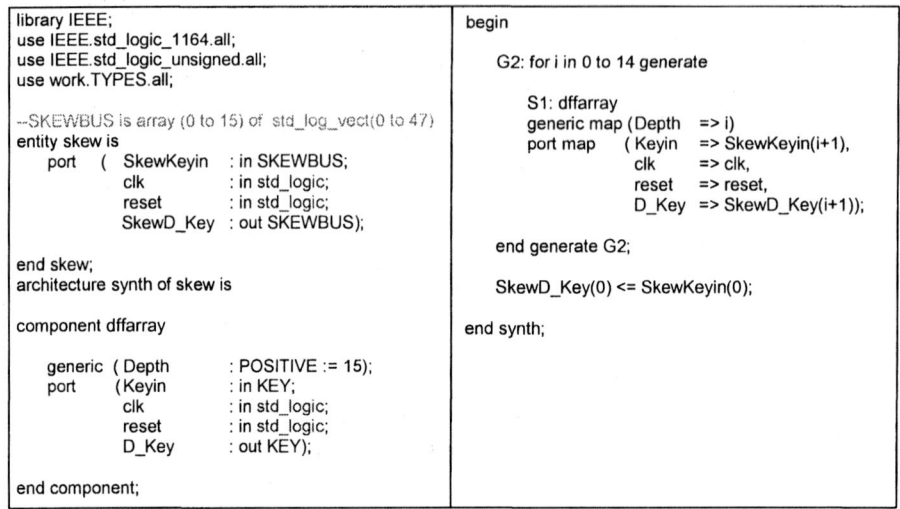

```
library IEEE;                                    begin
use IEEE.std_logic_1164.all;
use IEEE.std_logic_unsigned.all;                    G2: for i in 0 to 14 generate
use work.TYPES.all;
                                                     S1: dffarray
--SKEWBUS is array (0 to 15) of std_log_vect(0 to 47)  generic map (Depth  => i)
entity skew is                                         port map    ( Keyin  => SkewKeyin(i+1),
    port ( SkewKeyin  : in SKEWBUS;                                 clk     => clk,
           clk        : in std_logic;                               reset   => reset,
           reset      : in std_logic;                               D_Key   => SkewD_Key(i+1));
           SkewD_Key  : out SKEWBUS);
                                                    end generate G2;
end skew;
architecture synth of skew is                       SkewD_Key(0) <= SkewKeyin(0);

component dffarray                                end synth;

    generic ( Depth  : POSITIVE := 15);
    port    ( Keyin  : in KEY;
              clk    : in std_logic;
              reset  : in std_logic;
              D_Key  : out KEY);

end component;
```

Figure 2-18. Code for 'Skew' Component

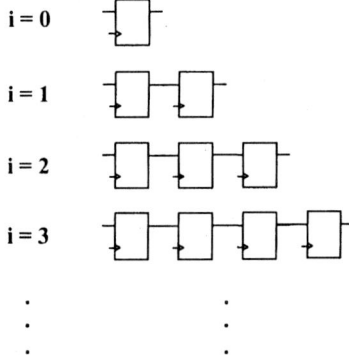

Figure 2-19. Array of Registers of Varying Lengths Generated by Skew Core

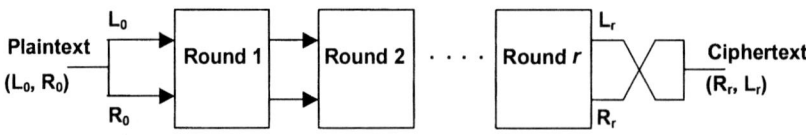

Figure 2-20. Fiestel Structure

DES Algorithm Architectures and Implementations

In algorithms similar to DES, where it is possible to obtain the sub-keys by performing permutations and substitutions, the key scheduling core will support the use of different keys every clock cycle. If it is not possible, the core may still be utilised provided one key is used per session. The Serpent algorithm, one of the five AES finalists, uses thirty-three 128-bit sub-keys. If Serpent's key generation block is designed and implemented, it will prove costly in terms of both area and speed. However, if one key is utilised throughout the data transfer session, the sub-keys may be pre-computed and hence loaded over thirty-three clock cycles prior to the start of the encryption/decryption process. In a pipelined version of the algorithm, the sub-keys could then be fed through the skew of registers to arrive at the required round at the correct time. Thus, utilising the skew design is a much more feasible method.

2.7.4. Performance Results

The pipelined DES design is a large design as it contains 16 instantiations of the function f component, hence for implementation purposes, the targeted FPGA device is the largest in the Virtex family, the XCV1000. Similar to the generic parameterisable DES core, distributed RAM are initialised and used to implement the S-Boxes. Eight 32 x 1-bit RAM blocks are required for each S-Box. The design is implemented using Xilinx Foundation Series software on the XCV1000-4BG560 device. Data blocks can be accepted every clock cycle and after an initial delay of 16 clock cycles the respective encrypted/decrypted data blocks appear on consecutive clock cycles. The design also supports a key change at full speed, i.e. a different key can be used with every block of plaintext.

The DES design incorporating the novel key scheduling method utilises 6446 CLB slices which is 52% of the total number of CLB slices available on this device. Of IOBs, 188 out of 404 (46%) are used. This design uses a system clock of 59.5 MHz and the data-rate achieved is 3.8 Gbits/sec [39]. It is possible to further enhance these performance figures by optimisation of the algorithm specific to the requirements of the FPGA device on which the design is implemented. However, this would result in the design being less easy to migrate to other devices and technologies.

2.8. Conclusions

DES has been a FIPS standard since 1977 and although it has recently been replaced, it will be required to retain compatibility with old products and it will remain as a benchmark for all new algorithms in the future.

A new generic parameterisable DES IP core architecture is described in this chapter. This architecture can generate DES designs to suit many application requirements. It accommodates feedback and non-feedback modes of operation, single or Triple-DES functionality, and shifting and permutation key scheduling techniques. It can also be utilised to create designs of specific speed and area configurations by varying the number of pipeline stages. For illustrative purposes, the designs that can be generated from the DES architecture are implemented on Virtex, Virtex-E and Virtex-II devices. However, technology-independent implementation of the DES S-Boxes is supported and therefore, the architecture is readily migratable to other FPGA/PLD and ASIC technologies.

The shifting method is the more efficient of the two key generation systems and proves suitable for designs requiring few key changes. Since the shifting method incurs an initial latency on inputting a new key during decryption, the permutation method is more appropriate for designs involving frequent key changes.

The 16-stage pipelined design is the most efficient of the pipelined designs. When implemented on a Virtex-E XCV300E device, the fully pipelined design achieves a throughput of 7.8 Gbits/sec. If implemented on a Virtex-II speed grade –8 device or indeed ASIC hardware, even higher data-rates can be achieved. The design is approximately 57 times faster than equivalent software implementations. It also compares very favourably with existing hardware FPGA implementations. Faster designs reported in the literature include that of Trimberger *et al.* [51], which achieves a throughput of 12 Gbits/sec and the design by Patterson [49], which operates at 10.7 Gbits/sec. However, in both these designs the key schedule is computed in software and can only support one key per data transfer session. The performance of the fully pipelined generic design is among the fastest hardware implementations currently available and is one of the fastest single-chip complete DES algorithm designs reported to date.

The performance of the iterative design is also comparable to similar previous hardware implementations. It runs at 278 Mbits/sec on a Virtex-II device and utilises 281 CLB slices. The only known faster iterative implementation is the core by CAST Inc. [54], which achieves a speed of 404 Mbits/sec.

A new key scheduling method for pipelined implementations of symmetric-key encryption algorithms is also presented in this chapter. It is a simple, easy-to-follow method, which involves the pre-computation and delayed presentation of the algorithm sub-keys. The DES algorithm is used to demonstrate the key scheduling technique in operation. The key scheduling method is generic and parameterisable and hence is migratable to any algorithm, which can be pipelined in its implementation. One conventional method of implementing the DES algorithm key schedule utilises logic cyclic shift operations at each stage of the 16-stage pipeline to create the sub-keys. The method of implementing the key schedule presented in §2.7 is an extension of the conventional permutation method and utilises permutations to create the sub-keys from the input key. The sub-keys are delayed by the required amount using the necessary array of registers. Hence the design allows the loading of a different key every clock cycle.

When implemented on a Virtex XCV1000 FPGA, the design achieves a throughput of 3.8 Gbits/sec. Similarly to the generic parameterisable DES designs, if implemented on enhanced Virtex devices or ASIC technology, higher throughputs can be obtained.

The NIST selected five finalists for the Advanced Encryption Standard in August 1999: MARS, RC6, Rijndael, Serpent and Twofish. MARS, RC6 and Twofish are fiestel-based algorithms while Rijndael and Serpent are substitution-permutation algorithms. The novel key scheduling design presented can also be utilised in a pipelined implementation of any of these algorithms.

Chapter 3

RIJNDAEL ARCHITECTURES AND IMPLEMENTATIONS

3.1. Introduction

On the 2nd October 2000 the US NIST selected the Rijndael algorithm, developed by Joan Daemen and Vincent Rijmen [56], as the new Advanced Encryption Standard (AES) algorithm. It proved a fast and efficient algorithm when implemented in both hardware and software across a range of platforms. In November 2001, the AES was approved as the Federal Information Processing Encryption Standard (FIPS 197) and it is to be employed by government agencies and the private sector to encrypt sensitive, unclassified information [57]. In the future Rijndael will be the encryption algorithm used in many applications such as:
- Internet Routers
- Remote Access Servers
- High Speed ATM/Ethernet Switching
- Satellite Communications
- Virtual Private Networks (VPNs)
- SONET
- Mobile phone applications
- Electronic Financial Transactions

This chapter describes high performance single-chip FPGA implementations of the Rijndael algorithm. To attain high throughputs, the designs are fully pipelined architectures. A fully pipelined Rijndael design requires considerable memory; hence, its implementation is ideally suited to the Virtex-E and Virtex-E Extended Memory range of FPGAs [24], which contain devices with up to 280 RAM Blocks (BRAMs). A novel generic parameterisable encryption-only architecture is described, from which designs can be generated to support 128-bit, 192-bit and 256-bit keys [58, 59]. The 128-bit key Rijndael encryption design, which achieves a throughput of 7 Gbits/sec, is one of the first and highest performance single-chip AES implementations reported in the literature [60]. The architecture also includes an efficient Rijndael key schedule that can be employed in both iterative and pipelined implementations [61]. A new Rijndael encryptor/decryptor architecture is discussed [59, 62]. It is one of the first fully pipelined Rijndael implementations capable of performing both encryption and decryption. Typically, pipelined implementations specifically support either encryption or decryption since to support both will result in a highly area-inefficient design. However, in the architecture outlined in this chapter, the similarities between the two operations are cleverly exploited and a high-throughput encryptor/decryptor design is achieved, while avoiding excess memory utilisation.

A detailed description of the Rijndael algorithm is provided in this chapter and a review of Rijndael hardware implementations is outlined. Performance evaluations for both the variable key AES design and encryption/decryption architecture are presented and a comparison provided with other Rijndael hardware implementations.

3.2. Rijndael Algorithm Description

The Rijndael algorithm is a substitution-linear transformation network [28]. It can operate on 128-bit, 192-bit and 256-bit data and key blocks. The NIST requested that the AES must implement a symmetric block cipher with a block size of 128 bits, hence the variations of Rijndael which can operate on larger data block sizes are not included in the actual FIPS standard. An outline of Rijndael is shown in Figure 3-1.

Rijndael comprises 10, 12 or 14 rounds when the key lengths are 128, 192 or 256 bits respectively. The transformations in Rijndael consider the data block as a four column rectangular array of 4-byte vectors (known as the *State* array). A 128-bit plaintext of 16-bytes, B_0, B_1, B_2, B_3 ... B_{15}, is represented as an array of four rows and four columns as illustrated in Figure 3-2. Similarly, the key is represented as a rectangular array of bytes, as in Figure 3-3, having four rows and a varying number of columns, N_k dependent on the key length. When the key length is 128, 192 or 256-bits, N_k is 4, 6 or 8 respectively.

Rijndael Algorithm Architectures and Implementations 59

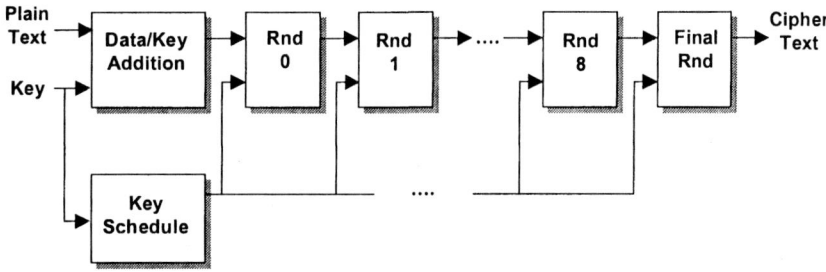

Figure 3-1. Outline of 128-bit Key Rijndael Encryption Algorithm

The algorithm consists of an initial data/key addition, nine, eleven or thirteen rounds when the key length is 128-bits, 192-bits or 256-bits respectively and a final round, which is a variation of the typical round. The Rijndael key schedule expands the key entering the cipher so that a different sub-key or round key is created for each algorithm iteration.

B_0	B_4	B_8	B_{12}
B_1	B_5	B_9	B_{13}
B_2	B_6	B_{10}	B_{14}
B_3	B_7	B_{11}	B_{15}

Figure 3-2. State Rectangular Array

		$N_k = 4$		$N_k = 6$		$N_k = 8$	
K_0	K_4	K_8	K_{12}	K_{16}	K_{20}	K_{24}	K_{28}
K_1	K_5	K_9	K_{13}	K_{17}	K_{21}	K_{25}	K_{29}
K_2	K_6	K_{10}	K_{14}	K_{18}	K_{22}	K_{26}	K_{30}
K_3	K_7	K_{11}	K_{15}	K_{19}	K_{23}	K_{27}	K_{31}

Figure 3-3. Key Rectangular Array

3.2.1. Rijndael Round

The Rijndael round, as outlined in Figure 3-4, consists of four transformations:
- SubBytes transformation   - MixColumns transformation
- ShiftRows transformation   - XorRoundKey transformation

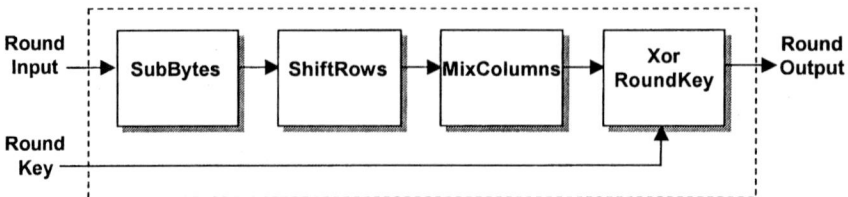

Figure 3-4. Rijndael Round

These transformations can be considered as a non-linear layer, a linear mixing layer and a key XOR layer. The SubBytes transformation is the *S-Box* of the Rijndael algorithm and it operates on each of the State bytes independently. It is the non-linear layer and is constructed by finding the multiplicative inverse of each byte in $GF(2^8)$ and then applying an affine transformation. The affine transformation involves multiplication by a matrix, outlined in equation (3.1), followed by addition to the hexadecimal number, 0x63.

$$\begin{bmatrix} b_0 \\ b_1 \\ b_2 \\ b_3 \\ b_4 \\ b_5 \\ b_6 \\ b_7 \end{bmatrix} = \begin{bmatrix} 1 & 0 & 0 & 0 & 1 & 1 & 1 & 1 \\ 1 & 1 & 0 & 0 & 0 & 1 & 1 & 1 \\ 1 & 1 & 1 & 0 & 0 & 0 & 1 & 1 \\ 1 & 1 & 1 & 1 & 0 & 0 & 0 & 1 \\ 1 & 1 & 1 & 1 & 1 & 0 & 0 & 0 \\ 0 & 1 & 1 & 1 & 1 & 1 & 0 & 0 \\ 0 & 0 & 1 & 1 & 1 & 1 & 1 & 0 \\ 0 & 0 & 0 & 1 & 1 & 1 & 1 & 1 \end{bmatrix} \begin{bmatrix} B_0 \\ B_1 \\ B_2 \\ B_3 \\ B_4 \\ B_5 \\ B_6 \\ B_7 \end{bmatrix} + \begin{bmatrix} 1 \\ 1 \\ 0 \\ 0 \\ 0 \\ 1 \\ 1 \\ 0 \end{bmatrix} \quad (3.1)$$

The linear mixing layer involves the ShiftRows transformation, in which the rows of the State are cyclically shifted to the left. Row 0 is not shifted, row 1 is shifted 2 places, row 2 is shifted 2 places and row 3 is shifted 3 places. It also includes column mixing using maximum distance separable (MDS) codes over $GF(2^8)$. The columns of the State are considered as polynomials over $GF(2^8)$ for the MixColumns transformation, and multiplied modulo $x^4 + 1$ with a fixed polynomial $c(x)$, where,

$$c(x) = \text{'03'} \, x^3 + \text{'01'} \, x^2 + \text{'01'} \, x + \text{'02'} \quad (3.2)$$

and '03', '01' and '02' are hexadecimal numbers. In the final round the MixColumns transformation is not included.

Rijndael Algorithm Architectures and Implementations

The XorRoundKey transformation is the key XOR layer, in which each byte in the round key is bitwise XORed to each byte in the State array. For example, a State array byte, B_0 will be XORed with the corresponding Round Key byte, RK_0, to obtain the output byte, A_0, as illustrated in Figure 3-5.

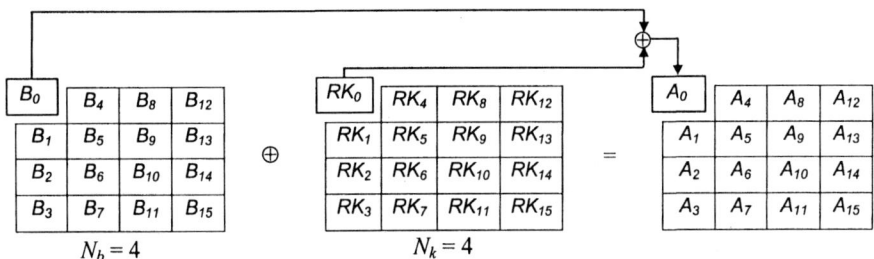

Figure 3-5. Round Key and State Array Addition

3.2.2. Rijndael Key Schedule

The Rijndael key schedule consists of two parts: Key Expansion and Round Key Selection. Key Expansion involves expanding the cipher key into a linear array of 4-byte words, the length of which is determined by the data block length, N_b, multiplied by the number of rounds, N_r plus 1, i.e. N_b * (N_r + 1). The data block length, N_b = 4. When the key block length, N_k = 4, 6 and 8, the number of rounds is 10, 12 and 14 respectively. Hence the lengths of the expanded key are as shown in Table 3-1.

Table 3-1. Length of Expanded Key for Varying Key Sizes

Data Block Length, N_b	4	4	4
Key Block Length, N_k	4	6	8
Number of Rounds, N_r	10	12	14
Expanded Key Length	*44*	*52*	*60*

For example, when using a 128-bit key, the number of rounds in the algorithm is 10 and therefore, the expanded key length is 44. The expanded key is a linear array of 4-byte words, W[0] to W[43], the first four words of which comprise the cipher key as illustrated in Figure 3-6.

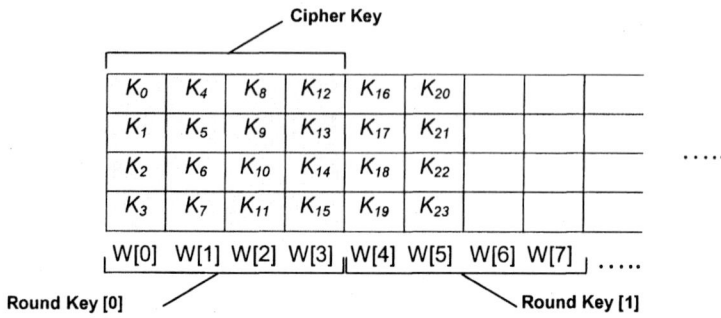

Figure 3-6. Expanded Key Array when Nk = 4

Each remaining word, W[i] is derived by XORing the previous word, W[i-1] with the word N_k positions earlier, W[i-N_k]. For words in positions, which are a multiple of N_k, a transformation is applied to W[i-1]. Firstly, the bytes in the word are cyclically shifted to the left. For example, a word [a,b,c,d] becomes [b,c,d,a]. Next, each byte in the word is passed through the Rijndael SubBytes transformation and finally, the result is XORed with a round constant. The round constants required for each of the rounds are 4-byte vectors, Rcon[i] = (RC[i], '00', '00', 00') where the values of RC[i] are as outlined in Table 3-2. However, when N_k = 8, an additional transformation is applied. For words in positions $i > 8$ where (i-4) is a multiple of N_k, each byte of the word, W[i-1], is passed through the Rijndael S-Box.

Table 3-2. Key Schedule Round Constants

RC[1] = '01'	RC[2] = '02'	RC[3] = '04'	RC[4] = '08'	RC[5] = '10'
RC[6] = '20'	RC[7] = '40'	RC[8] = '80'	RC[9] = '1B'	RC[10] = '36'

In the Round Key selection process, the round keys are extracted from the expanded key. In a design with N_r rounds, N_r +1 round keys are required. For example a 10-round design requires 11 round keys. Round key 0 is W[0] to W[3] and is utilised in the initial data/key addition, round key 1 is W[4] to W[7] and is used in round 0, round key 2 is W[8] to W[11] and used in round 1 and so on as shown in Figure 3-6. Finally, round key 10 is used in the final round.

3.2.3. Decryption

The decryption process in Rijndael is effectively the inverse of its encryption process. Data is first passed through an inverse of the final round, then the inverses of the rounds and finally through the initial data/key addition. The data/key addition remains the same as it involves an XOR operation, which is its own inverse. The inverse of the round is found by inverting each of the transformations in the round. The inverse of the SubBytes operation is obtained by applying the inverse of the affine transformation and taking the multiplicative inverse in $GF(2^8)$ of the result. In the inverse of the ShiftRows transformation, row 0 is not shifted, row 1 is now shifted 3 places, row 2 by 2 places and row 3 by 1 place. Similar to the data/key addition, Round Key addition is its own inverse. To invert the MixColumns transformation, the State columns are multiplied modulo $x^4 + 1$ with a fixed polynomial $d(x)$, where,

$$d(x) = \text{'0B'}x^3 + \text{'0D'}x^2 + \text{'09'}x + \text{'0E'} \qquad (3.3)$$

and '0B', '0D', '09' and '0E' are hexadecimal numbers.

During decryption, the key schedule does not change, however the round keys constructed are now used in reverse order. For example, in a 10-round design, round key 0 is utilised in the initial data/key addition and round key 10 in the inverse of the final round. Round key 1 is used in the inverse of round 8, round key 2 in the inverse of round 7 and so on.

3.3. Review of Rijndael Hardware Implementations

Recently, a lot of attention has been focused on hardware implementations of encryption algorithms as they achieve much higher data rates than software-only solutions. This has been motivated by the growth of technologies such as broadband wireless communications. Since the selection of Rijndael as the AES, the highest performance single-chip Rijndael design and one of the first implementations is the variable key design presented in §3.4. [58, 59]. This is based on a single-chip Xilinx Virtex-E implementation. As will be discussed, the 7 Gbit/sec design is a 128-bit key fully pipelined encryptor core.

Other work on AES algorithm hardware implementations has included very high throughput designs, and small, low area designs. Currently, the fastest Rijndael FPGA implementation reported in the literature is a heavily pipelined design by Chodowiec *et al.* [63], which achieves a throughput of 12160 Mbits/sec. However, this requires three Xilinx Virtex XCV1000 devices. It performs both encryption and decryption but only supports a 128-

bit key and non-feedback modes of operation. A 5-stage pipelined design by Elbirt *et al.* [17] on the same device performs at a data-rate of 1937.9 Mbits/sec. McMillan and Patterson [64] carried out a Jbits implementation of AES on an XCV1000 device, achieving a throughput of 900 Mbits/sec utilising only 288 CLB slices and 32 BRAMs. However, similar to the DES architecture by Patterson [49], the key schedule is performed in software and thus the design is not a single-chip implementation of the full Rijndael algorithm.

A number of ASIC implementations of the Rijndael algorithm also exist. Ichikawa, Kasuya and Matsui's [33] unrolled implementation in 0.35μm CMOS operates at 1950 Mbits/sec. The encryptor/decryptor implementation by Weeks *et al.* [32] in 0.5μm CMOS, performs at a rate of 5745 Mbits/sec.

Table 3-3 provides a specification summary of the 128-bit key Rijndael pipelined and unrolled implementations described above.

Table 3-3. Specifications of 128-bit Rijndael Pipelined and Loop Unrolled Implementations

Manufacturer	Type of Design	Device Used	Area	Data Rate Mbits/sec
Chodowiec, Khuon, Gaj [63] *Over 3 devices*	P	XCV1000 x3	12600 CLB slices 80 BRAMs	12160
Elbirt *et al.* [17]	SP	XCV1000	10992 CLB slices	1938
McMillan, Patterson [64] *Jbits implementation*	P	XCV1000	288 CLB slices 32 BRAMs	900
Weeks *et al.* [32]	P	0.5μm CMOS	420 mm^2	5745
Ichikawa *et al.* [33]	UL	0.35μm CMOS	612,000 gates	1950

Iterative designs of the AES include a 414 Mbit/sec design also by Chodowiec *et al.* [63] and a 294 Mbit/sec design by Elbirt *et al.* [17], which utilises 3528 CLB slices on a XCV1000 device. In both designs only a 128-bit key is supported and the key scheduling is performed off-chip. A 353 Mbit/sec design by Dandalis, Prasanna and Rolim [23] incorporates the key schedule, however, decryption, feedback modes of operation and longer key lengths are not supported. The Jbit iterative design by McMillan and Patterson [64] again implements the key schedule in software and achieves a throughput of 250 Mbits/sec. The ASIC iterative implementation by Weeks *et al.* [32] operates at 606 Mbits/sec.

Rijndael Algorithm Architectures and Implementations

The majority of Rijndael implementations have used Xilinx Virtex devices with some being implemented on CMOS ASICs. One exception is the work of Mroczlowski [65], which is based on an Altera EPF10K250 PLD. This achieves a data-rate of 268 Mbits/sec utilising 20 Embedded Array Blocks (EABs) and 1032 Logic Cells (LCs).

Table 3-4 summarises these 128-bit key Rijndael iterative hardware implementations. The fastest Rijndael software implementation is Gladman's [66] 325 Mbit/sec design on a 933 MHz Pentium III processor.

Table 3-4. Specifications of 128-bit Key Rijndael Iterative Implementations

Manufacturer	Type of Design	Device Used	Area	Data Rate Mbits/sec
Chodowiec, Khuon, Gaj [63]	IL	XCV1000	2507 CLB slices	414
Dandalis et al. [23]	IL	XCV1000	5673 CLB slices	353
Elbirt et al. [17]	IL	XCV1000	3528 CLB slices	294
McMillan, Patterson [64] *Jbits implementation*	IL	XCV1000	240 CLB slices 8 BRAMs	250
Mroczlowski [65]	IL	Altera EPF10k250	1032 LCs 20 EABs	268
Weeks et al. [32]	IL	0.5μm CMOS	34 mm^2	606

3.4. Design of High Speed Rijndael Encryptor Core

The Rijndael algorithm implementations presented in this chapter concentrate on achieving high throughputs and thus are fully pipelined implementations. Both designs are based on the Electronic Codebook (ECB) mode of operation. Since they are fully pipelined implementations they will also operate in Counter mode as discussed in chapter 2. The NIST's recommended Rijndael modes of operation are outlined in chapter 4.

The main consideration in both the encryptor architecture and the 128-bit key encryptor/decryptor design described in §3.5, is the memory requirement. The Rijndael S-Box in the SubBytes transformation can be implemented as a look-up table (LUT) or ROM. The values contained in this LUT are given in Appendix C.1. This proves a faster and more cost-effective method than implementing the multiplicative inverse operation and affine transformation. Since the State bytes are operated on individually, each Rijndael round requires sixteen 8-bit to 8-bit LUTs. In the key schedule, LUTs can also be used, as words are passed through the S-Box. The Virtex-E and Virtex-E Extended Memory range of FPGAs are utilised for implementation as they contain devices with up to 280 BRAMs.

A single BRAM can be configured into two single port 256 x 8-bit RAMs, as illustrated in Figure 3-7; hence, 8 BRAMs are required for each round. As described in §2.6.4, when the write enable of the RAM is low, transitions on the write clock are ignored and data stored in the RAM is not affected. Hence, if the RAM is initialised and both the input data and write enable pins are held low then the RAM can be utilised as a ROM or LUT.

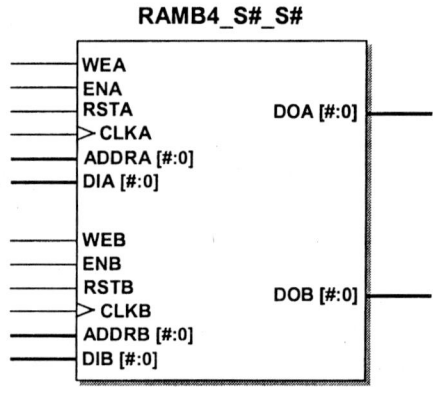

Figure 3-7. Dual-Port Block SelectRAM

The ShiftRows transformation is simply hardwired as no logic is involved. The MixColumns transformation can be written as a matrix multiplication as given in equation (3.4), with a 4-byte input, a_0, a_1, a_2, a_3 and output, b_0, b_1, b_2, b_3.

$$\begin{bmatrix} b_0 \\ b_1 \\ b_2 \\ b_3 \end{bmatrix} = \begin{bmatrix} 02 & 03 & 01 & 01 \\ 01 & 02 & 03 & 01 \\ 01 & 01 & 02 & 03 \\ 03 & 01 & 01 & 02 \end{bmatrix} \begin{bmatrix} a_0 \\ a_1 \\ a_2 \\ a_3 \end{bmatrix} \tag{3.4}$$

The transformation is implemented by XORing the results of the multiplications in $GF(2^8)$ in accordance with equation (4.4), as illustrated in Figure 3-8.

Rijndael Algorithm Architectures and Implementations 67

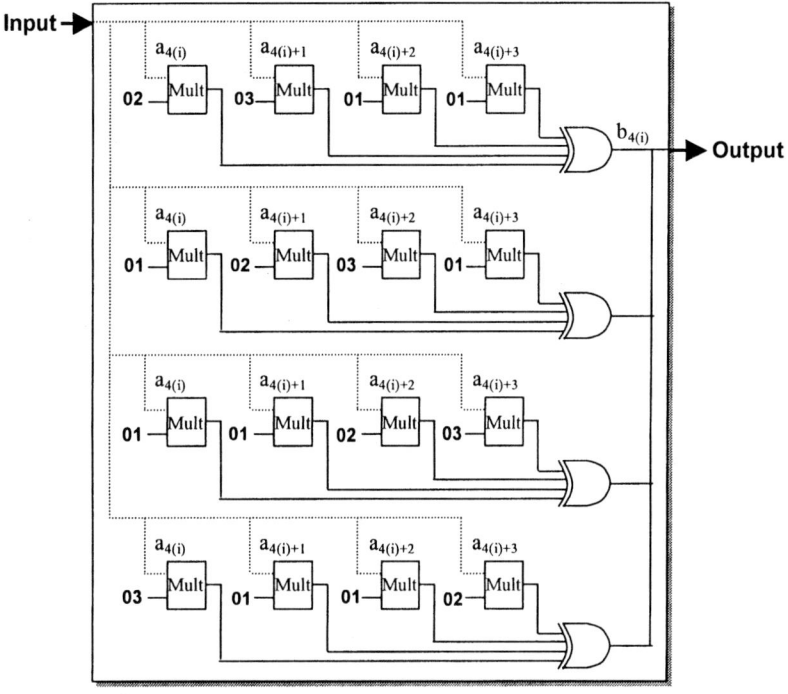

Figure 3-8. Design of MixColumns Transformation

3.4.1. Design to Support Three Key Lengths

The key scheduling component and the number of rounds are affected by a requirement to support the varying key lengths. The 128, 192 and 256-bit key lengths require 10, 12 and 14 rounds respectively. Thus, an architecture that supports 14 rounds and all three key lengths is initially designed. In the VHDL code, *generate* statements are used to select the logic required for each key length. This means that if a key length of 128-bits is required, only the logic for that particular key length will be synthesised and similarly for the 192 and 256-bit key lengths. Hence, two extra rounds will only be created if a 192-bit key is required and four extra rounds will only be created when a 256-bit key is selected.

3.4.2. Key Schedule Design

The flowchart in Figure 3-9 outlines the various stages involved in the Rijndael key schedule for key lengths of 128, 192 and 256-bits. N_k, N_b and N_r represent the key block length, the data block length and the number of rounds respectively. The input to the key schedule is the cipher key and key block length and the outputs are the round keys. The round keys are created as required, hence, round key [0] should be available immediately, round key [1] should appear one clock cycle later and so on.

The various functions utilised in the key schedule are as follows:

Rem Function: Returns the remainder value in a division. For example, 12/8 = 1 remainder 4; therefore, 12 rem 8 = 4.

SubWord Function: Operates on a 4-byte word and each byte is passed through the Rijndael S-Box.

RotWord Function: Involves a cyclic shift to the left of the bytes in a 4-byte word. For example, an input of x_0, x_1, x_2, x_3, will produce the output x_1, x_2, x_3, x_0.

Rcon Function: Returns the 4-byte round constants outlined in Table 3-2.

When utilising a 128-bit key, forty words are created during key expansion and every fourth word is passed through the SubBytes transformation with each byte in the word being transformed. Hence, forty 8-bit to 8-bit LUTs or twenty BRAMs can be utilised in its implementation. However, since the round keys are constructed in parallel to the round operations, only two BRAMs are required. The S-Box is only used in the construction of the first word of every round key (a round key comprises four words) and each BRAM is used in the construction of two bytes of a word.

Therefore an iterative process can be used to access the two BRAMs and the round keys are constructed as they are required by each Rijndael round. The design assumes that the same key is used in any one data transfer session.

The construction of every fourth word, $i = 4, 8 \ldots 40$, which incorporates the BRAMs, is shown in Figure 3-10. As described in §3.2.2, words which are not a multiple of four are created by XORing the previous word with the word four positions earlier.

Rijndael Algorithm Architectures and Implementations

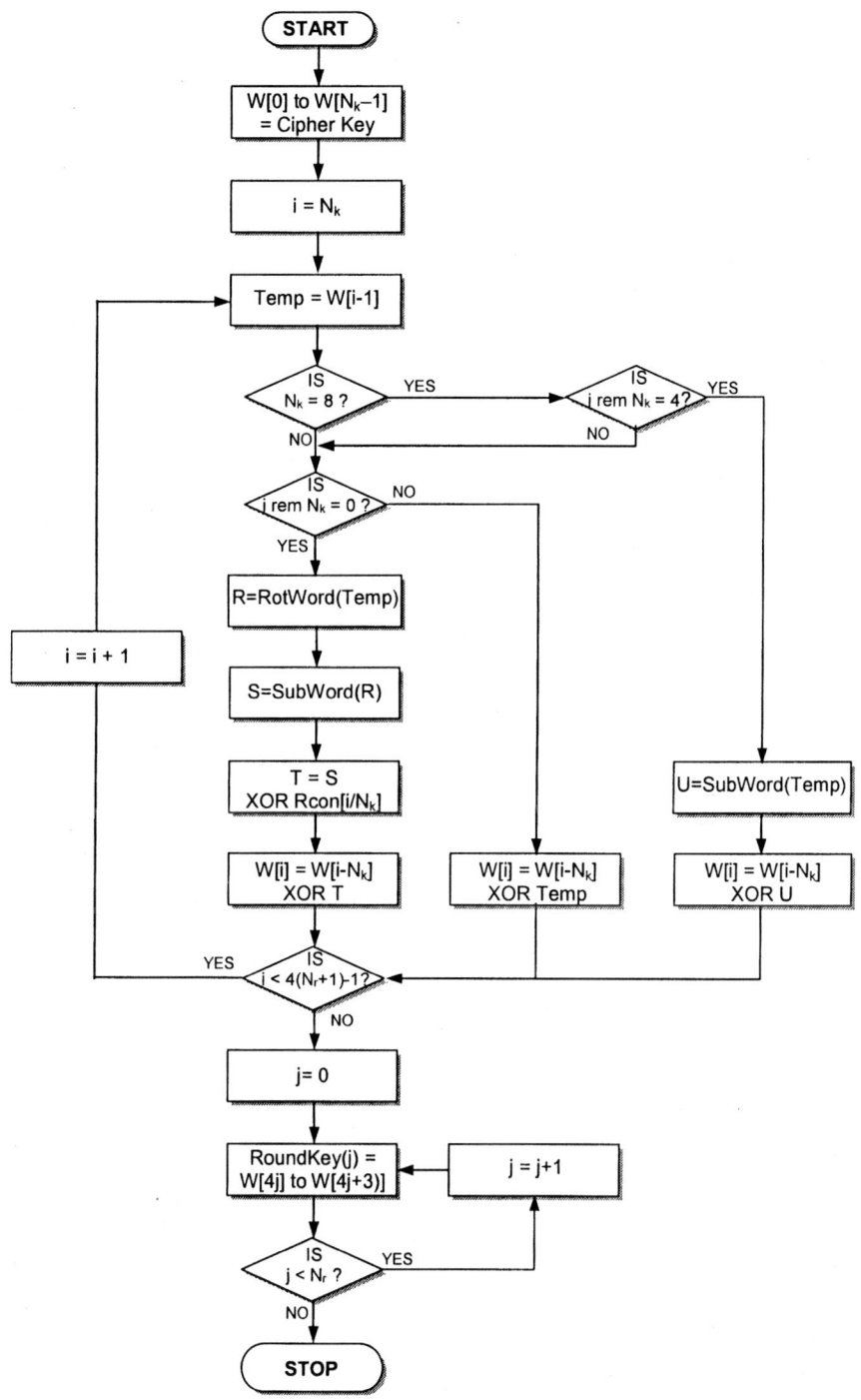

Figure 3-9. Rijndael Key Schedule

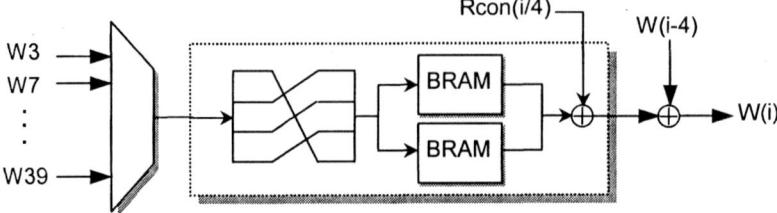

Figure 3-10. Construction of every Fourth Word in 128-bit Rijndael Key Schedule

Therefore, in the fully pipelined 128-bit key Rijndael design, a total of 82 BRAMs are utilised – 80 BRAMs are required for the 10 rounds and a further 2 for the key schedule. In the 192-bit architecture every sixth word is passed through the SubBytes transformation and hence, the 192-bit key schedule also requires 2 BRAMs. Similarly, in the 256-bit key design every eighth word and each word $i > 8$, where $(i-4)$ is a multiple of 8 requires the use of 2 BRAMs. Therefore, 98 and 114 BRAMs are required by these implementations respectively.

3.5. Encryptor/Decryptor Core

In Rijndael decryption, the inverse of the SubBytes transformation can also be implemented as a LUT. However the values in this LUT are different to those required for encryption. The LUT required in Rijndael decryption is given in Appendix C.2. Therefore, it is necessary to include both LUTs in order to accommodate encryption and decryption. One method would involve doubling the number of BRAMs utilised, however, this would prove costly on area. In the novel Rijndael encryption/decryption design presented here, this was overcome by the addition of just two BRAMs, which were utilised as ROMs, one containing the initialisation values for the LUTs required during encryption, the other containing the values for the LUTs required during decryption. Therefore, instead of initialising each individual BRAM as a ROM, when the design is set to encrypt, all the BRAMs are initialised with data read from the ROM containing the values required for encryption. When the design is set to decrypt, the BRAMs are initialised with data from the ROM containing the values required for the decryption operation. This initialisation procedure is outlined in Figure 3-11. Effectively, only one initialising BRAM is necessary to store the required encryption and decryption values. However, since such a high number of BRAMs need to be initialised, the use of two BRAMs helps reduce overall fanout.

Rijndael Algorithm Architectures and Implementations

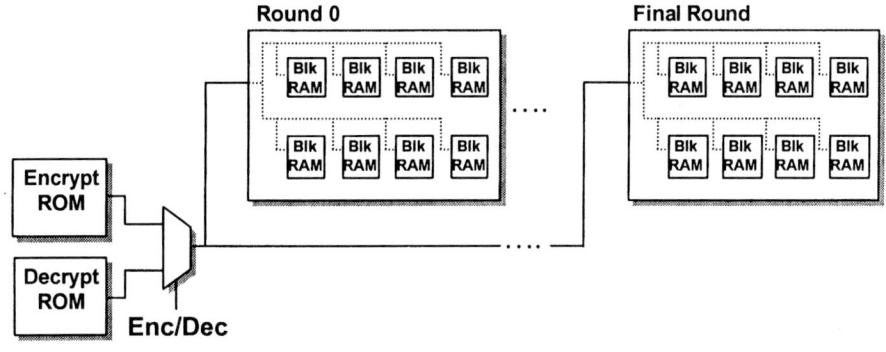

Figure 3-11. Initialisation of Block RAMs in Rijndael Design

The InvShiftRows transformation is hardwired. Multiplexors (MUXs) select between the ShiftRows and InvShiftRows wiring as depicted in Figure 3-12. Similarly to Figure 3-8, the InvMixColumn transformation can be implemented by XORing results of the multiplications in $GF(2^8)$ arising from equation (3.5). Again, MUXs are used to select between the values required for encryption and those required during decryption.

$$\begin{bmatrix} b_0 \\ b_1 \\ b_2 \\ b_3 \end{bmatrix} = \begin{bmatrix} 0E & 0B & 0D & 09 \\ 09 & 0E & 0B & 0D \\ 0D & 09 & 0E & 0B \\ 0B & 0D & 09 & 0E \end{bmatrix} \begin{bmatrix} a_0 \\ a_1 \\ a_2 \\ a_3 \end{bmatrix} \quad (3.5)$$

Since the encryptor/decryptor core design assumes a key length of 128 bits, the design of the key schedule is a simplification of that shown in the flowchart illustrated in Figure 3-9. During decryption, the values of the LUTs utilised in the key schedule do not change, hence, the LUTs can simply be implemented as ROMs. However, the round keys are used in reverse order. The initialisation process for either encryption or decryption takes 256 clock cycles as the 256 values contained in each ROM are read. However, this delay can be reduced by the addition of further initialising ROMs. When encrypting data, the keys are produced as each round requires them, therefore, the encryption will take 10 clock cycles corresponding to the 10 rounds when using a 128-bit key. The design assumes that the same key is utilised during a session of data transfer. If decrypting data, the initialisation process will be as described above. However, initial decryption will take 20 clock cycles, 10 clock cycles for the required round keys to be constructed and a further 10 corresponding to the 10 rounds. The overall 128-bit key encryptor/decryptor design, therefore, requires 84 BRAMs.

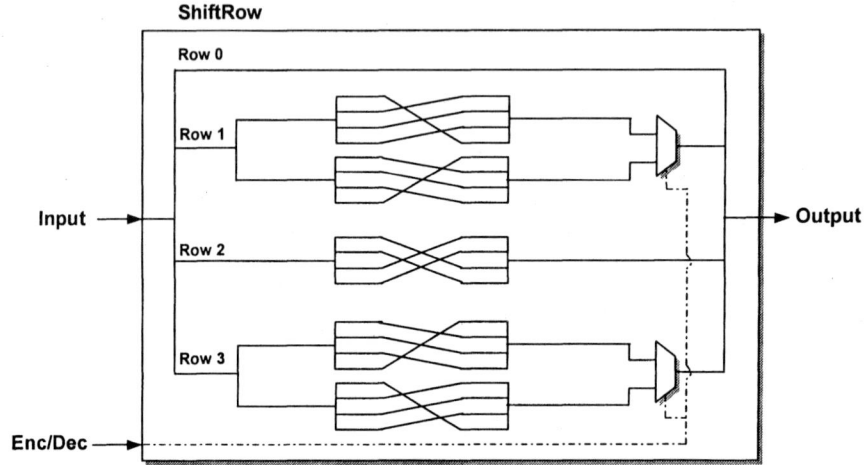

Figure 3-12. ShiftRows and InvShiftRows in Encryptor/Decryptor Design

3.6. Performance Results

The Rijndael designs were implemented using Xilinx Foundation Series 3.1i software and Synplify Pro V6.0 on Xilinx Virtex-E FPGA devices. Data blocks can be accepted every clock cycle and after an initial delay the respective encrypted/decrypted data blocks appear on consecutive clock cycles. The designs were verified using the test vectors outlined in the Rijndael specification [56].

The 128-bit key Rijndael encryptor design, implemented on the Virtex-E XCV812E-8BG560 device, utilises 2679 CLB slices (28%) and 82 BRAMs (29%). Of the available IOBs, 385 of 404 are used. The design can use a maximum system clock of 54.35 MHz and runs at a data-rate of 7 Gbits/sec. This result proves to be one of the fastest single-chip Rijndael FPGA implementations currently available. As discussed earlier, the only faster FPGA implementation is that of Chodowiec, Khuon and Gaj [63], which has a throughput of 12.16 Gbit/sec. However, this design requires 3 Virtex XCV1000 devices.

For comparison purposes a Rijndael 128-bit decryptor design was also implemented on the XCV812E-8BG560 FPGA device. This utilises 4304 slices (45%) and 82 BRAMs. It operates at a rate of 6.38 Gbit/sec using a maximum system clock of 49.9 MHz. The variance in the encryption and decryption performances is due to the different multiplier constants required by each design. In encryption the multiplier constants are simply 0x01, 0x01,

0x02 and 0x03 (hexadecimal) while those used in decryption mode are 0x0B, 0x0E, 0x09 and 0x0D.

Both the 192-bit and 256-bit key encryption designs have been implemented on Virtex-E XCV3200E-8-CG1156 devices, as they require a higher number of IOBs than that available on the XCV812E device. The 192-bit key encryption design utilises 2577 CLB slices (7%) and 98 BRAMs. Of IOBs 448 of 804 are used. The design uses a maximum system clock of 45.44 MHz and runs at a data-rate of 5.8 Gbits/sec. The 256-bit key architecture requires 2995 CLB slices (9%), 114 BRAMs and 512 IOBs. The design operates at 39.88 MHz with a 5.1 Gbit/sec throughput. Other FPGA implementations of Rijndael designs capable of supporting 128-bit, 192-bit and 256-bit key lengths have typically been iterative designs. Weeks *et al.* [32] carried out a fully pipelined ASIC implementation, which performs at 5.3 Gbits/sec for all three key lengths. Hence the 128-bit and 192-bit key encryption designs prove more efficient than this ASIC implementation, while the 256-bit key architecture achieves a comparable speed.

The Rijndael encryptor/decryptor architecture has also been implemented on the Virtex-E XCV3200E-8-CG1156 device. This implementation utilises 7576 CLB slices (23%), 84 BRAMs and 385 IOBs. It runs at 25.3 MHz and achieves a data-rate of 3.24 Gbits/sec. Since the initialisation process for either encrypting or decrypting data is 256 clock cycles and the system clock is 25.3 MHz, this corresponds to an initialisation time of 10 µs. Other encryptor/decryptor FPGA implementations have also typically been iterative designs. However, Ichikawa *et al.*'s [33] fully unrolled 1950 Mbit/sec ASIC design and the 5745 Mbit/sec ASIC implementation by Weeks *et al.* [32] perform both encryption and decryption. Therefore, the design presented here compares very well with existing ASIC implementations.

Table 3-5 summarises the performance results of both the high-speed generic variable key encryptor design, the 128-bit key decryptor implementation and the novel encryptor/decryptor implementation.

The high performance of the Rijndael designs presented is achieved for a number of reasons:
- The designs are fully pipelined with data blocks being accepted on every clock cycle.
- The use of dedicated Block RAMs: The complex and slow operations involved in the SubBytes transformation, the multiplicative inverse calculations over $GF(2^8)$ and matrix multiplication and addition, are replaced with simple LUTs.
- The layout of the Virtex-E architecture: An outline of the Virtex-E architecture is provided in Figure 3-13 [24]. It is evident that the BRAMs are located in columns throughout the

chip, with each memory column extending the full height of the chip. Each Rijndael round involves implementation on both CLBs and BRAMs. Therefore, having an architecture where these are located in close vicinity to one another throughout the chip improves overall performance.

Table 3-5. Summary of High-Speed Rijndael Pipelined Implementations

Design	Device	Area (CLB slices)	No. of BRAMs	Throughput (Mbits/sec)
Generic Encryptor Core: 128-bit Key	XCV812E	2679	82	6956
Generic Encryptor Core: 192-bit Key	XCV3200E	2577	98	5816
Generic Encryptor Core: 256-bit Key	XCV3200E	2995	116	5104
128-bit Key Decryptor Core	XCV812E	4304	82	6387
128-bit Key Encryptor/ Decryptor Core	XCV3200E	7576	84	3239

3.7. Conclusions

High performance single-chip FPGA implementations of the Rijndael algorithm are presented in this chapter. The generic variable key encryptor architecture is among the fastest designs currently available which is capable of generating cores to support all three required key lengths. The 128-bit key encryption design performs at a data-rate of 7 Gbits/sec, which is 3.6 times faster than similar existing FPGA implementations and 21 times faster than software implementations. An efficient Rijndael key schedule design that can be utilised in both iterative and pipelined designs is also discussed.

Many previous Rijndael encryption-only designs are implemented on Virtex XCV1000 devices, which consist of only 32 BRAMs and therefore, cannot support a fully pipelined Rijndael design. The Virtex-E and Virtex-E Extended Memory family of FPGAs, however, contain up to 280 BRAMs and can easily accommodate large unrolled designs. When used to implement the Rijndael designs, these devices help to improve overall throughput.

Rijndael Algorithm Architectures and Implementations

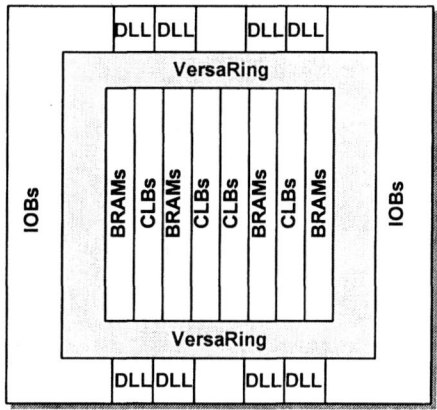

Figure 3-13. Virtex-E Architecture Overview

The novel encryptor/decryptor core runs at 3.2 Gbits/sec. This implementation compares favourably with similar ASIC designs but is one of the only fully pipelined high-speed single-chip FPGA Rijndael designs capable of both encryption and decryption. Typically, fully pipelined designs that support both modes incur extremely high area overheads. However, the architecture presented here achieves a high throughput while avoiding excess RAM usage. This is afforded by the simple addition of two ROMs, which are used to initialise the BRAMs required in each round with the respective encryption or decryption values.

Rijndael has been approved by NIST as the FIPS encryption standard and is set to replace DES in applications such as IPSec protocols, the Secure Socket Layer (SSL) protocol and in ATM cell encryption. In general, hardware implementations of encryption algorithms and their associated key schedules are physically secure, as they cannot easily be modified by an outside attacker. Also, the high speed Rijndael encryptor core and Rijndael encryptor/decryptor core presented, should prove beneficial in applications where speed is vital as with real-time communications such as satellite communications, SONET OC-48 networks and electronic financial transactions.

Chapter 4

FURTHER RIJNDAEL ALGORITHM ARCHITECTURES AND IMPLEMENTATIONS

4.1. Introduction

Since the selection of the AES in 2000, many Rijndael algorithm hardware implementations have been carried out with emphasis on achieving either low area or high performance designs. This is evident in the previous chapter. In this chapter, the designs described in chapter 3 are further developed and two novel Rijndael architectures are presented.

In the first of these architectures, a Look-Up Table (LUT) based methodology is introduced, whereby the complex operations of the Rijndael algorithm – the multiplicative inverse operation and multiplication and addition in $GF(2^8)$ – are replaced by LUTs. Therefore, these operations are, in effect, pre-computed and the expected results for all possible inputs placed in LUTs. This approach leads to high area requirements, however, it produces very fast implementations. To illustrate this, a 10-stage fully pipelined LUT based Rijndael encryptor design [62, 67, 68] is described. However, due to the area requirements, the LUT approach is best suited to smaller iterative Rijndael designs.

An alternative generic and migratable architecture is also presented [69]. This allows the instantiation of a wide range of chip specifications. Cores implemented from this architecture can perform both encryption and decryption and support four modes of operation - Electronic Codebook (ECB) mode, Output Feedback (OFB) mode, Cipher Block Chaining (CBC) mode and Ciphertext Feedback (CFB) mode. Chip designs can also be generated to cover all three AES key lengths, 128-bits, 192-bits and 256-bits.

On-the-fly generation of the round keys required during decryption is also possible [70]. The general, flexible and multi-functional nature of the approach described, contrasts with previous designs which, to date, have been focused on specific implementations. For the purposes of this chapter the ideas presented are demonstrated by implementation in FPGA technology. However, the architecture and IP cores derived from this are easily migratable to other silicon technologies, including ASIC and PLD, and are capable of covering a wide range of modern communication systems cryptographic requirements.

This chapter begins with a description of the LUT-based Rijndael designs. Next, the modes of operation which can be used with Rijndael are discussed and the generic, flexible AES cryptographic architecture is then presented. Performance evaluations and comparisons for both architectures are also provided.

4.2. Look-Up Table Based Rijndael Architecture

In the LUT-based design approach described in this chapter, the complex and slow operations of the Rijndael algorithm are implemented utilising LUTs. These operations, which include multiplicative inverse operations and multiplication and addition in $GF(2^8)$, can be implemented in logic. However, in the approach described here, these operations are pre-computed for all possible inputs and the results placed in LUTs. High-speed designs can be achieved utilising the LUT method and this is shown through the implementation of a fully pipelined Rijndael encryptor core. The LUT methodology, however, leads to high silicon area utilisation. Therefore, the approach is evidently better suited to smaller area iterative designs. Hence, LUT-based encryptor and decryptor iterative designs are also presented.

4.2.1. Fully Pipelined LUT-Based Design

In the LUT-based Rijndael designs it is evident that the SubBytes transformation or *S-Box* of the Round function can be implemented as a LUT or ROM. This proves a much more efficient method than implementing the multiplicative inverse operation and affine transformation. However, the ShiftRows and MixColumns transformations can also be implemented as LUTs rather than using logic. The design is based on equation (4.1),

$$e_j = T_0[a_{0,j}] \oplus T_1[a_{1,j+1}] \oplus T_2[a_{2,j+2}] \oplus T_3[a_{3,j+3}] \oplus k_j \qquad (4.1)$$

where *a* and *k* represent the state and key inputs to the round respectively, and *e* represents the output, as shown in Figure 4-1.

Further Rijndael Algorithm Architectures and Implementations

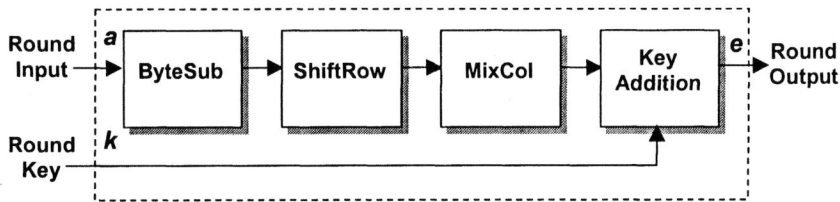

Figure 4-1. Rijndael Round During Encryption

The functions T_0, T_1, T_2, and T_3 are outlined in equation (4.2).

$$T_0 = \begin{bmatrix} S[a] \bullet 02 \\ S[a] \\ S[a] \\ S[a] \bullet 03 \end{bmatrix} \quad T_1 = \begin{bmatrix} S[a] \bullet 03 \\ S[a] \bullet 02 \\ S[a] \\ S[a] \end{bmatrix} \quad T_2 = \begin{bmatrix} S[a] \\ S[a] \bullet 03 \\ S[a] \bullet 02 \\ S[a] \end{bmatrix} \quad T_3 = \begin{bmatrix} S[a] \\ S[a] \\ S[a] \bullet 03 \\ S[a] \bullet 02 \end{bmatrix} \quad (4.2)$$

(*S[a]* represents a byte, *a*, being operated on by the Rijndael S-Box)

Therefore, two further LUTs are required: LUT_02, containing the values of the SubBytes LUT multiplied in $GF(2^8)$ by the hexadecimal number 0x02, i.e. *S[a]• 02*, and LUT_03, containing values of the SubBytes LUT multiplied in $GF(2^8)$ by the hexadecimal number 0x03, i.e. *S[a]• 03*. LUT_02 and LUT_03 are given in Appendix D.1. Since the State bytes are operated on individually, each Rijndael round will require 48 8-bit to 8-bit LUTs or 24 BRAMs. Figure 4-2 illustrates the design required to achieve the first two Round outputs. The final round excludes the MixColumns transformation and hence only needs 8 BRAMs.

The key schedule has been designed using the iterative approach outlined in §3.4.2 and thus only requires 2 BRAMs in its implementation. Therefore, the overall pipelined design utilises 226 BRAMs (216 utilised in the 9 typical rounds, 8 in the final round and 2 for the key schedule).

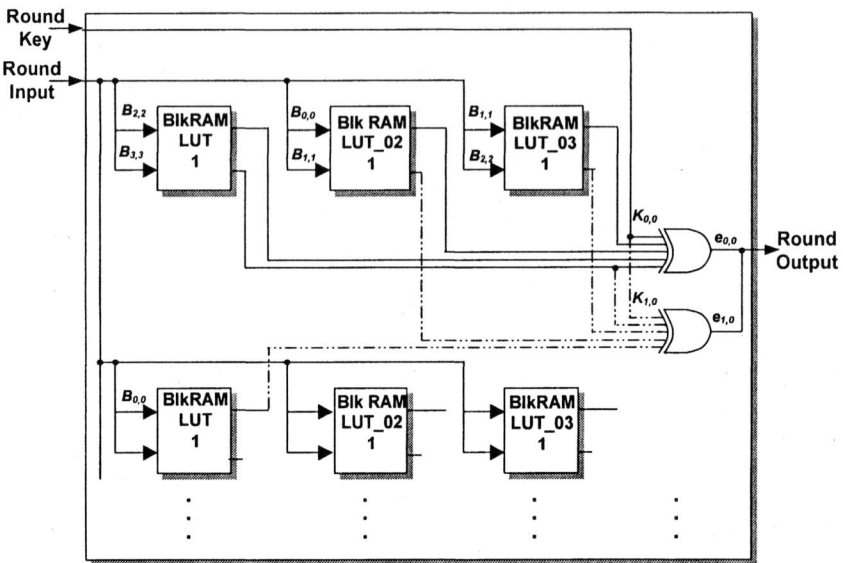

Figure 4-2. LUT Based Rijndael Round Design During Encryption

In the Rijndael round during decryption, as illustrated in Figure 4-3, the data block and inverse round key are XORed and the result passed through the InvMixColumns transformation, the InvShiftRows and the InvSubBytes transformation. The design of a decryption-only implementation is based on equation (4.3) where, b is the output of the data/key addition and e is the output of the InvSubBytes transformation. Five LUTs are required. The first LUT is the inverse of the LUT utilised in the SubBytes transformation, *InvLUT*. The four other LUTs required are $(0E \bullet b_{i,j})$, $(0B \bullet b_{i,j})$, $(0D \bullet b_{i,j})$ and $(09 \bullet b_{i,j})$.

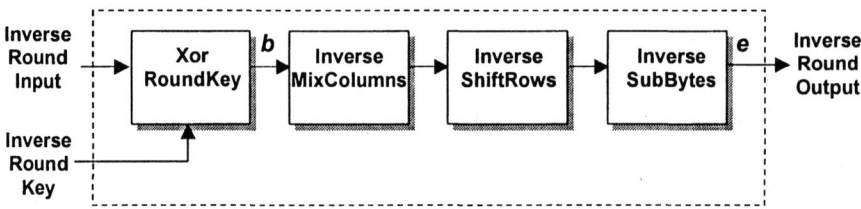

Figure 4-3. Rijndael Round During Decryption

Further Rijndael Algorithm Architectures and Implementations 81

$$\begin{bmatrix} e_{0,j} \\ e_{1,j} \\ e_{2,j} \\ e_{3,j} \end{bmatrix} = \begin{bmatrix} InvLUT[0E \bullet b_{0,j} \oplus 0B \bullet b_{1,j} \oplus 0D \bullet b_{2,j} \oplus 09 \bullet b_{3,j}] \\ InvLUT[09 \bullet b_{0,j+3} \oplus 0E \bullet b_{1,j+3} \oplus 0B \bullet b_{2,j+3} \oplus 0D \bullet b_{3,j+3}] \\ InvLUT[0D \bullet b_{0,j+2} \oplus 09 \bullet b_{1,j+2} \oplus 0E \bullet b_{2,j+2} \oplus 0B \bullet b_{3,j+2}] \\ InvLUT[0B \bullet b_{0,j+1} \oplus 0D \bullet b_{1,j+1} \oplus 09 \bullet b_{2,j+1} \oplus 0E \bullet b_{3,j+1}] \end{bmatrix} \quad (4.3)$$

For example, the LUT ($0E \bullet b_{i,j}$) is constructed by multiplying every possible byte from 0x00 to 0x11 by 0x0E in GF(2^8) and similarly for the other LUTs. These LUTs are outlined in Appendix D.2. 40 BRAMs per round are required to implement a decryption-only design. The circuit required to achieve the first two Round outputs during decryption is shown in Figure 4-4.

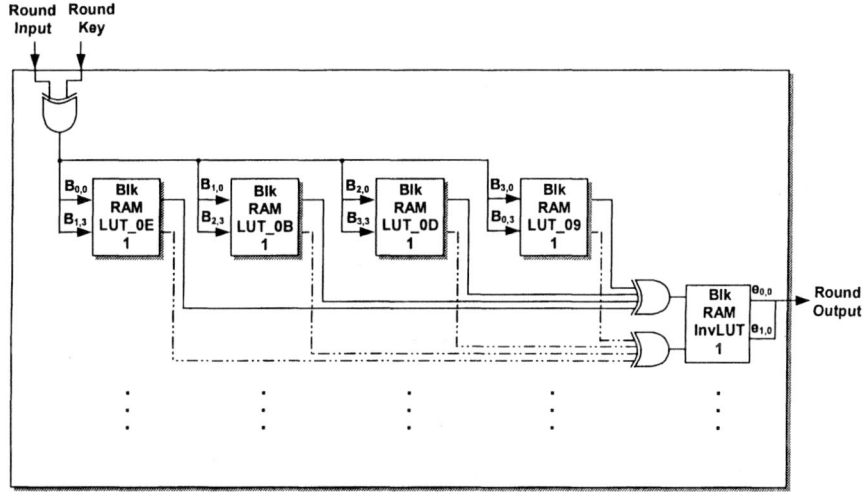

Figure 4-4. LUT Based Rijndael Round Design During Decryption

The inverse of the final round can be implemented using only 8 BRAMs since the InvMixColumns transformation is excluded. The key schedule utilises 2 BRAMs, hence an overall decryptor design requires 370 Block RAMs. It is not possible to implement a fully pipelined decryptor design on an FPGA as yet, since the highest number of Block RAMs incorporated in a device (XCV812E FPGA) is 280. In June 2002, Xilinx released an advance product specification specifying Virtex-II Pro FPGA devices, which will contain up to 556 BRAMs [71]. Therefore, it will be possible to implement the LUT-based decryptor design in the near future.

4.2.2. Iterative LUT-Based Design

The LUT-based design approach is best suited to iterative implementations which are naturally low in area. In the look-up table based iterative Rijndael implementation, the Round is designed similarly to that outlined above in §4.2.1 whereby the SubBytes, ShiftRows and MixColumns transformations are all implemented using LUTs. A block diagram of the overall iterative encryptor Rijndael design is shown in Figure 4-5. In this design the plaintext is loaded over 4 clock cycles in 32-bit blocks and similarly the ciphertext appears in 32-bit blocks. This leads to a lower IOB count. There is an initial latency of 14 clock cycles before the first 128-bit block of encrypted data appears; 4 clock cycles corresponding to the plaintext loading and 10 clock cycles corresponding to the 10 rounds. Subsequent encryption operations take 10 clock cycles as plaintext blocks can be pre-loaded during the last four cycles of the Rijndael round encryption operation. The key schedule is designed using the iterative method, with each Round key being created as it is required by a Rijndael Round. Control circuitry is used to synchronise the timing of the overall design.

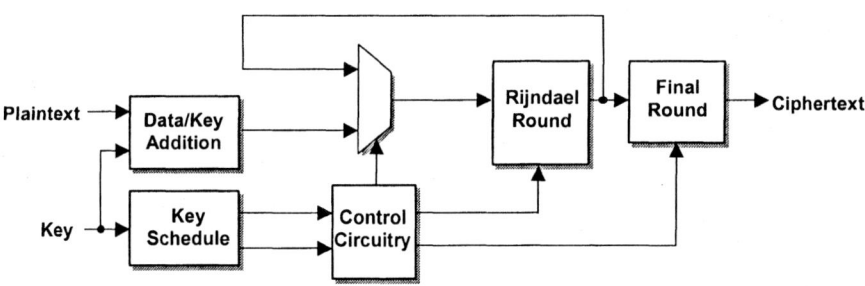

Figure 4-5. Overall Iterative Rijndael Block Diagram During Encryption

The Rijndael iterative encryptor design utilises 34 BRAMs – the Round requires 24 BRAMs, the Final Round requires 8 BRAMS and the key schedule uses 2 BRAMs.

An iterative decryptor design is possible using the LUT method. The block diagram for the decryptor Rijndael design is depicted in Figure 4-6. The LUT-based iterative decryptor design requires 50 BRAMs in its implementation. The Final Round again only requires 8 BRAMs and the key schedule 2 BRAMs. The Inverse Round utilises 40 BRAMs. An inverse round takes 2 clock cycles to perform since each State byte must pass though 2 LUTs – either *LUT_0B, LUT_0E, LUT_0D or LUT_09* and the *InvLUT*. However, 2 blocks of data can be operated on at any one time and therefore, after 20 clock cycles two 32-bit plaintext blocks will appear over 8 clock cycles at the output. Hence, during decryption, the initial latency is 30 clock cycles. It is necessary to wait 10 clock cycles for the Round keys to be created and a further 20 clock cycles corresponding to the rounds. The ciphertext input blocks can be loaded while the Round keys are being created. Subsequent decryptions will take 20 clock cycles per 2 blocks of data.

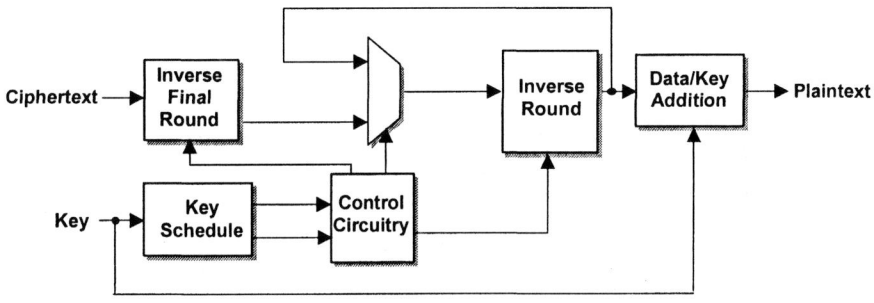

Figure 4-6. Overall Iterative Rijndael Block Diagram During Decryption

4.2.3. Performance Results and Comparison

The Rijndael designs presented have been implemented using Xilinx Foundation Series 3.1i software and Synplify Pro V6.0 on Xilinx Virtex-E FPGA devices. Synplify Pro V6.0 has been used to provide pre-placement timing results for the 128-bit key Rijndael encryptor design. The fully pipelined LUT-based design implemented on the XCV812E-8-BG560 device utilises 2457 of 9408 CLB slices (26%) and 226 of 280 BRAMs (81%). The implementation runs at a system clock of 93.9 MHz and achieves a data-rate of 12 Gbits/sec. In comparison, timing results from pre-placement of the design described in §3.4, which uses BRAMs in the implementation of the SubBytes transformation only, indicate a throughput of 9.2 Gbits/sec. Although CLB usage in the pipelined design is low, many applications have additional circuitry which can utilise the remaining CLB resources. Therefore, in terms of pre-placement timing, the design is the fastest single-chip Rijndael FPGA implementation currently available. However, in order to obtain a corresponding high post-placement speed, manual placement and routing of the design is required due to the large number of BRAMs.

The LUT-based iterative encryptor and decryptor designs presented have been implemented on the XCV400E-8-BG432 and XCV600E-8-BG432 devices respectively. The encryption design requires 1926 of 4800 CLB slices and 34 of 40 BRAMs. It achieves a throughput of 685 Mbits/sec. The decryption design uses 1937 of 6912 CLB slices, 50 of 72 BRAMs and operates at 675 Mbits/sec. Both encryption and decryption LUT-based implementations display a performance that exceeds that of previously reported circuits. For example, the most competitive alternative implementation is that reported by Weeks *et al.* [32]. This operates at 606 Mbit/sec and is based on a 0.5μm CMOS ASIC implementation.

Table 4-1 summarises the performance results of the LUT-based designs. These results clearly highlight the benefits of the approach developed as a result of this research.

Table 4-1. Summary of LUT-Based Rijndael Implementations

Design	Type of Design	Device	Area (CLB slices)	No. of BRAMs	Through-put (Mbits/sec)
128-bit key LUT-Based Encryptor Core (Pre-placement timing)	P	XCV812E	2457	226	12020
128-bit key LUT-Based Encryptor Core	IL	XCV400E	1926	34	685
128-bit key LUT-Based Decryptor Core	IL	XCV600E	1937	50	675

4.3. Rijndael Modes of Operation

The FIPS Publication 81 [41] defines various modes of operation for the DES algorithm, which may be used in a wide variety of applications. However, the Rijndael algorithm has replaced DES as the FIPS encryption standard. Therefore, in conjunction with the approval of AES, the NIST issued a special publication 800-38A [72] in December 2001, in which they recommended five confidentiality modes for use with any approved block cipher algorithm. In this publication the four DES modes of operation, ECB, CBC, OFB and CFB, have been updated and the Counter mode (CTR) is added. The Rijndael ECB, CBC and CFB modes are similar to the equivalent DES modes of operation, as outlined in chapter 2. The main difference in the updated standard is that any approved symmetric algorithm can be utilised in the modes and not solely the DES algorithm. In this section, the modes are

Further Rijndael Algorithm Architectures and Implementations 85

described with the assumption that Rijndael is the underlying encryption algorithm.

4.3.1. ECB Mode

Rijndael operating in ECB mode involves 128-bit blocks of data being encrypted by the Rijndael algorithm to obtain 128-bit blocks of ciphertext. The ciphertext blocks are then sent to the recipient of the message where they are decrypted to give the original data.

4.3.2. CFB Mode

In CFB mode the Initialisation Vector (*IV*) block is encrypted utilising the Rijndael algorithm and n-bits of the result are XORed with n-bits of the plaintext to produce n-bits of ciphertext. The n-bit ciphertext is fed back to form part of the input to be used in the next encryption. After the first encryption the initial input is shifted left by n-bits and the next input therefore consists of a *(128 – n)*-bit data block concatenated with the n-bits of the ciphertext. Decryption in CFB mode also uses Rijndael encryption. However, in decryption n-bits of the result are XORed with n-bits of the ciphertext to produce n-bits of the original plaintext.

4.3.3. CBC Mode

The *IV* block is XORed with the first 128-bit plaintext in CBC mode, and the result encrypted using Rijndael to obtain the first ciphertext block. The next plaintext block is XORed with the first ciphertext block prior to encryption. This process is continued until a new message is loaded. CBC mode in decryption uses Rijndael decryption and the *IV* blocks are XORed with the decryption results to obtain the original plaintext.

4.3.4. OFB Mode

OFB mode is different to its equivalent DES mode in that the operations are carried out on full 128-bit blocks of data rather than n-bits of a data block. The *IV* is encrypted utilising the Rijndael algorithm and the result is XORed with the plaintext to produce the ciphertext. The result is then fed back to form the input used in the next encryption as shown in Figure 4-7. The *IV* data must be different for each encryption operation carried out using a specific key. If the final plaintext block is only a partial block of *y*-bits, it is XORed with the most significant *y*-bits of the final encryption result to obtain the corresponding ciphertext. Decryption in OFB mode uses Rijndael encryption. During decryption, however, the result is XORed with the ciphertext to produce the original plaintext.

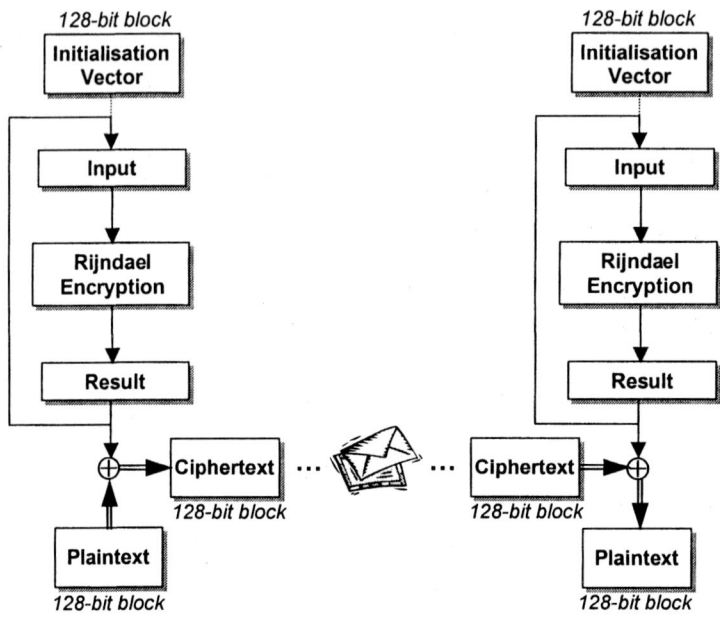

Figure 4-7. Rijndael in OFB Mode

4.3.5. Counter (CTR) Mode

In Counter (CTR) mode a counter is used in place of the *IV* block. Therefore, a 128-bit counter block is encrypted using Rijndael and the output is XORed with a plaintext block to obtain the ciphertext, as illustrated in Figure 4-8. It is essential that a different counter value is used for each new plaintext block. Similar to OFB and CFB mode, decryption requires the Rijndael encryption process. The same counter block is encrypted, but in decryption the result is XORed with the ciphertext to obtain the original plaintext block.

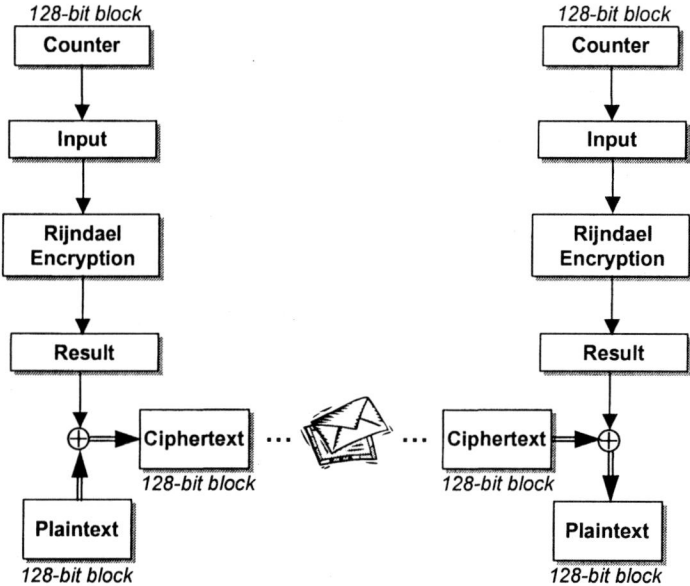

Figure 4-8. Rijndael in CTR Mode

4.4. Overall Generic AES Architecture

Previous AES implementations have been specific-purpose solutions. The objective of the research described in this section was to achieve an efficient, generic, flexible and silicon migratable AES architecture which contained the best combination of features and characteristics for use in modern communication applications. In order to achieve such an implementation a number of aspects have to be considered.

Performance

The performance of the implementation can be crucial. The encryption algorithm design must accommodate communication transmission rates. Slow running cryptographic algorithms translate into consumer dissatisfaction while at the other extreme, fast running encryption might mean high product costs since very high speed systems typically imply custom-built hardware devices [73].

Area

The area utilised by an implementation is also an issue – very high speed cores can be achieved by adopting heavily pipelined designs. However, this leads to very high silicon area, which implies high product costs and high power consumption.

Functionality

Many cryptographic algorithm implementations are specific one-off designs. They are designed solely to either encrypt or decrypt data with specifically defined parameters. However, the majority of applications require a core capable of performing more than one function. The NIST requested that the AES algorithm would accept not one, but three key lengths. Also a design which performs both encryption and decryption is an obvious requirement for many applications.

Reusability

In recent years, with a shift to the system-on-chip integration, it is important to consider *design reuse* so that cores can be integrated quickly into silicon systems [74]. In general, Hardware Description Languages (HDLs) provide methods to capture reusable designs without compromising the integrity of the underlying designs [75]. VHDL features that facilitate reuse include generics, packages, generate statements and configuration specifications. A wholly generic design can facilitate a varying number of pipeline stages. However, if pipeline stages are accommodated, feedback modes of operation cannot be supported. Also, only very specific applications would require fully pipelined encryption designs, which achieve high speeds in the range of gigabits per second at the expense of very high area utilisation.

It is very important to achieve the correct balance between design performance, area, functionality and reusability in any implementation. The overall generic architecture presented in this chapter attempts to combine all these characteristics in an optimal way. The design is an iterative architecture, which leads to lower area utilisation and higher computational efficiency. This patented approach [62, 70] can generically accept three different key lengths, 128-bits, 192-bits and 256-bits with on-chip key scheduling. The design supports both encryption and decryption and can also be operated in ECB mode and the feedback modes, CBC, OFB and CFB. Counter mode is not incorporated in the design as it is not yet supported in many applications. However, modifications to include the mode can be easily achieved since it is a simplification of OFB mode [29]. The memory blocks in the design are targeted towards Xilinx FPGA Virtex Block RAMs. However, these components are easily modified into technology-independent memories, and thus reusable components, as discussed in §4.4.4.

4.4.1. Design of Generic AES Architecture

The generic Rijndael architecture is an iterative design. The 128-bit data and the 128-bit *IV* block required in the feedback modes are loaded in 32-bit blocks over 4 clock cycles. Similarly, the 128-bit, 192-bit and 256-bit keys are entered in 32-bit blocks over 4, 6 and 8 clock cycles respectively. This ensures that the number of input and output pins required on the chip is kept to a minimal, and thus, smaller and cheaper hardware devices can be chosen for implementation. This proves highly attractive to both FPGA and ASIC technologies. When using a 128-bit key, data blocks are accepted every 10 clock cycles – the data is entered over four clock cycles and encryption is performed in the next 10, corresponding to the 10 algorithm rounds. In the 192-bit and 256-bit key designs, data is accepted every 12 and 14 clock cycles respectively. Similar to the input data, the encrypted/decrypted data is output in 32-bit blocks over 4 clock cycles. Many applications will only require the use of one of the three possible key sizes. Therefore, the key length is a generic input of the design. Having all three key length design options available only increases the area overhead unnecessarily. However, if a different key size is required, the core does not need to be re-designed – the generic key-length input value is simply changed and the design re-synthesised.

Both encryption and decryption are supported. The main distinction between encryption and decryption-only designs is that the values of the LUTs used to implement the SubBytes transformation differ. One method to provide both encryption and decryption capabilities involves the addition of 2 initialising LUTs or ROMs, as described in §3.5. However, this method incurs a latency each time there is a change of mode from encryption to decryption and vice versa. In this generic architecture, the change from encryption to decryption of data can occur on-the-fly. Since the architecture is iterative, the Round and Inverse Round transformations can both be included, incurring only a small area overhead. Figure 4-9 pictorially describes the encrypt/decrypt Rijndael design for a key length of 128-bits. The *load counter* loads the data and key over 4 cycles in 32-bit blocks. Similarly, the *output counter* outputs the encrypted/decrypted data over 4 cycles. The *128-Key counter* controls the timing of the overall design. The 192-bit and 256-bit key implementations are similar to that shown in Figure 4-9. The only difference between these designs is in the load and key counters. In these cases the load counter loads the key over 6 and 8 cycles respectively and the data over 4 cycles. 192-key and 256-key counters are also required to control the timing in these circuits.

The Rijndael round is as described in previous architectures. The SubBytes transformation has been implemented using 16 8-bit to 8-bit LUTs or 8 BRAMs when this is targeted towards a Xilinx Virtex device. Similarly the InvSubBytes transformation in the inverse round has been implemented utilising 16 LUTs. An iterative process has been used to access the round or inverse round for nine, eleven or thirteen cycles depending on the key length, analogous to the algorithm's 9, 11 or 13 typical rounds. On the tenth, twelfth or fourteenth cycle, again depending on the key length, the MixColumns and InvMixColumns transformations are bypassed corresponding to the final round and inverse final round respectively. The key schedule for encryption is designed using the iterative process outlined in §3.4.2. The 128-bit, 192-bit and 256-bit key schedule designs each require 2 BRAMs in their implementation.

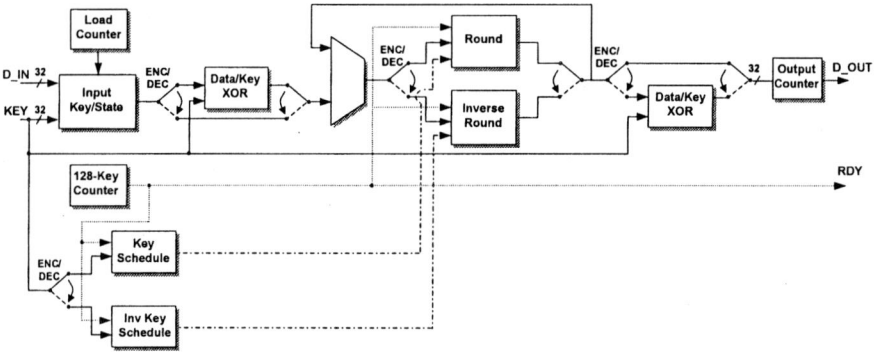

Figure 4-9. Outline of Encrypt/Decrypt Rijndael 128-bit Key Design

4.4.2. On-the-Fly Generation of Decryption Rounds Keys

During decryption, the Rijndael key schedule remains unchanged. The round keys created during key expansion are simply utilised in reverse order as depicted in Figure 4-10. Typically, during the decryption process it is therefore necessary to wait at least 10, 12 or 14 clock cycles, depending on the key length, for the round keys to be created before decryption can commence. Registers can be used to store the keys until they are required. Alternatively, the Round keys can be pre-computed and stored in memory.

However, on-the-fly calculation of the Round Keys for decryption can be achieved. If the final N_k words created during key expansion in the encryption process are utilised as the Cipher Key during decryption, the Round keys required for decryption can be created as they are needed by the inverse Rounds.

Further Rijndael Algorithm Architectures and Implementations 91

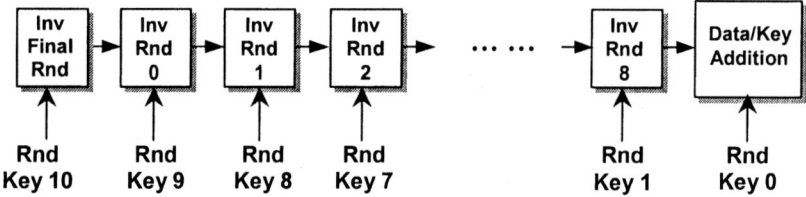

Figure 4-10. Round Key Utilisation in Decryption Process for 128-bit Cipher Key

```
when Nk = 4 or 6:
    for ( i = Nb*(Nr+1) -1; i ≥ Nb*(Nr+1) – Nk ; i -- )
        W[i] = (  InvCipherKey[4*(Nb*(Nr + 1)-1-i)], InvCipherKey[4*(Nb*(Nr + 1)-1-i)+1],
                  InvCipherKey[4*(Nb*(Nr + 1)-1-i)+2], InvCipherKey[4*(Nb*(Nr + 1)-1-i)+3);

    for ( i = Nb*(Nr+1) -1; i ≥ Nk; i -- )
    {
        temp = W[i - 1];
        if (i % Nk == 0)
            temp = SubWord(RotWord(temp)) XOR Rcon(i/Nk);
        W[i-Nk] = W[i] XOR temp;
    }

when Nk = 8:
    for ( i = Nb*(Nr+1) -1; i ≥ Nb*(Nr+1) – Nk ; i -- )
        W[i] = (  InvCipherKey[4*(Nb*(Nr + 1)-1-i)], InvCipherKey[4*(Nb*(Nr + 1)-1-i)+1],
                  InvCipherKey[4*(Nb*(Nr + 1)-1-i)+2], InvCipherKey[4*(Nb*(Nr + 1)-1-i)+3);

    for ( i = Nb*(Nr+1) -1; i ≥ Nk; i -- )
    {
        temp = W[i - 1];
        if (i % Nk == 0)
            temp = SubWord (RotWord (temp)) XOR Rcon(i/Nk);
        else if (i % Nk == 4)
            temp = SubWord (temp);
        W[i-Nk] = W[i] XOR temp;
    }
```

Figure 4-11. Pseudo-Code of the Inverse Key Expansion Procedure

The N_k words used as the Cipher Key in decryption are known as the 'Inverse Cipher Key'. Typically, for decryption, the sender of the ciphertext will send the receiver the original key used to encrypt the message. However, if the sender sends the receiver the Inverse Cipher Key, the receiver can then begin to decrypt the message immediately. Hence, the 10, 12 or 14 clock cycle latency is removed and there is no need to store Round keys in memory. Pseudo-code for the inverse key expansion procedure is outlined in Figure 4-11, where N_k is the key block length, N_b is the data block length and N_r is the number of rounds. Block diagrams illustrating the design of the memory-less decryption key scheduler for the three key lengths are provided in Figure 4-12, Figure 4-13 and Figure 4-14.

The 128-bit, 192-bit and 256-bit key designs require a total of 20 BRAMs (40 LUTs) – 8 in the round, 8 in the inverse round, 2 in the encryption key schedule and 2 in the decryption key schedule.

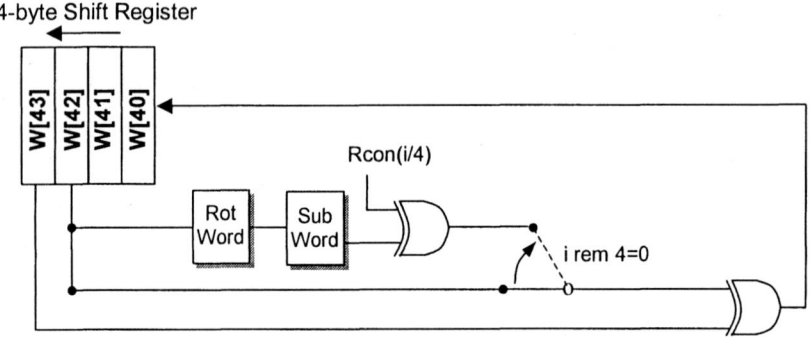

Figure 4-12. Hardware Design of Inverse Key Scheduler: Nk = 4

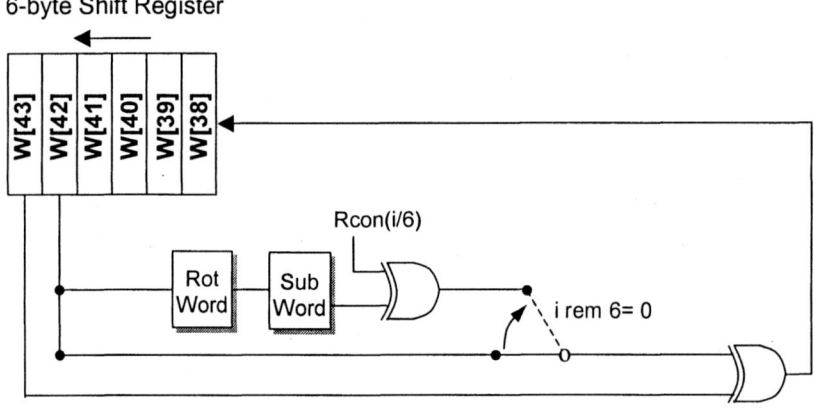

Figure 4-13. Hardware Design of Inverse Key Scheduler: Nk = 6

4.4.3. Design to Support Modes of Operation

Since the Rijndael design is iterative, it is possible to support the various feedback modes of operation. Figure 4-15 illustrates an outline of an architecture that includes the ECB, CBC, OFB and CFB modes of operation.

In ECB mode the input data is directly entered into the Rijndael core. Encryption or decryption is performed and the result is the output data. Data blocks can be accepted every 10, 12 or 14 clock cycles.

Further Rijndael Algorithm Architectures and Implementations 93

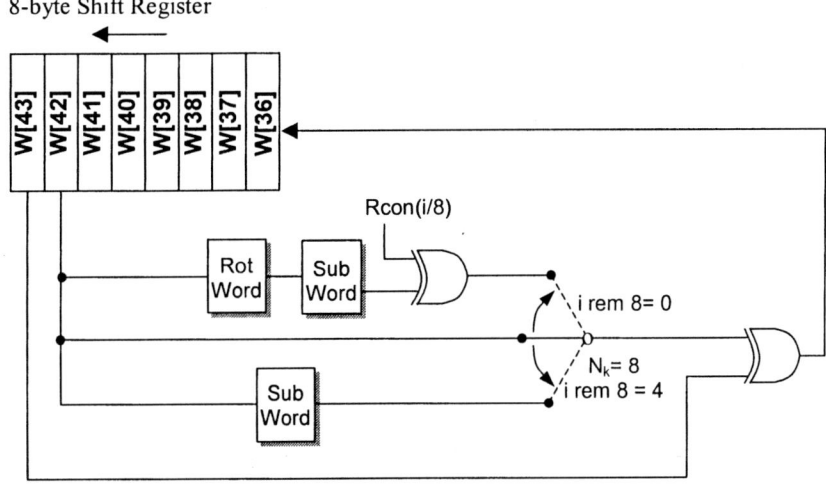

Figure 4-14. Hardware Design of Inverse Key Scheduler: Nk = 8

In the two feedback modes, CFB and OFB, the 128-bit initialisation vector is entered into the algorithm core and Rijndael encryption performed. The result is then XORed with the input data to produce the CFB or OFB output. In OFB mode for both encryption and decryption, the Rijndael result is fed back to form the next input of the algorithm core. During encryption in CFB mode the output ciphertext is fed back, while during decryption, the input ciphertext is fed back.

To encrypt data in CBC mode, the input data is initially XORed with the initialisation vector and the result is entered into the Rijndael core. Encryption is performed and the result is the output data. The encrypted result is then fed back and XORed with the next input data block. To decrypt data, the input data is directly entered into the algorithm core. Decryption is performed and the result is XORed with the initialisation vector to form the output. The ciphertext input will be XORed with the next decrypted input block result to form the next output.

The feedback modes operate on 128-bit plaintext blocks, hence, similarly to ECB mode, new data blocks can be accepted every 10, 12 or 14 clock cycles. Data transmission rates differ between applications. Therefore, to accommodate these varying transmission rates, a FIFO could be used to control the flow of the input data to the cores.

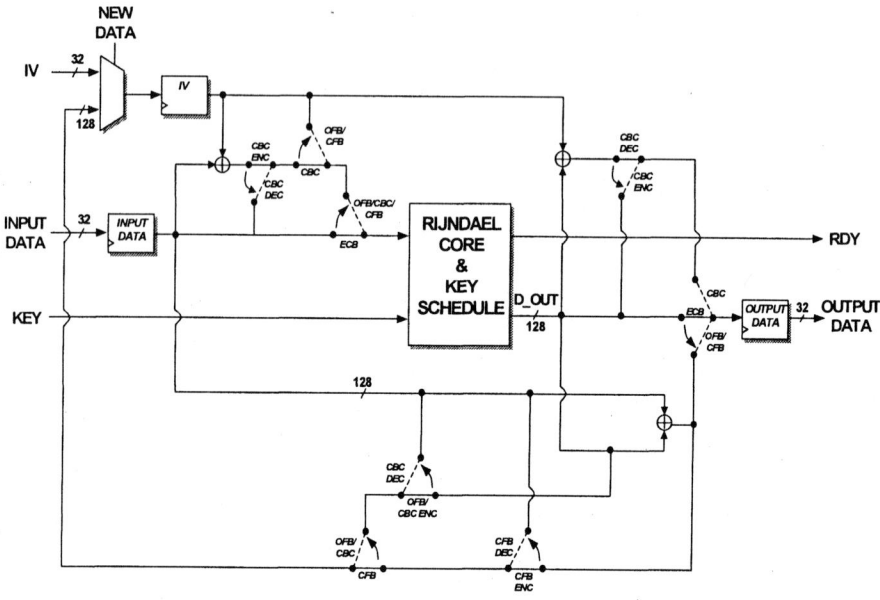

Figure 4-15. Outline of Design for Modes of Operation

4.4.4. Reusable Memory Components

The S-Boxes of the Rijndael algorithm considered to this point have been implemented as LUTs, targeted towards Xilinx Virtex-E Block RAM components. However, it is possible to create general, technology-independent memory banks. For example, a small 2-bit to 2-bit LUT can be described in VHDL as a general ROM component as illustrated in Figure 4-16. To target the general ROM component to a Virtex Block RAM component simply involves adding an attribute to the code. For example, if the design is synthesised using Synplify Pro the attribute *syn_ramstyle* can be utilised, as shown in Figure 4-16. The 2-bit to 2-bit LUT is then mapped to a Virtex RAMB4_S2 component. The RAMB4_S2 component can also be directly instantiated in VHDL, as outlined in Figure 4-17. However, this limits the use of the design to only Xilinx Virtex devices, whereas the general memory design is reusable in other technologies.

```
library ieee;
use ieee.std_logic_1164.all;
use ieee.std_logic_unsigned.all;

library synplify;
use synplify.attributes.all;

entity smallmem is
            port(   clk,write,reset : in std_logic;
                    Addr            : in std_logic_vector(10 downto 0);
                    DataIn          : in std_logic_vector(1 downto 0);
                    DataOut         : out std_logic_vector(1 downto 0));
end smallmem;

architecture smallmem_synth of smallmem is

type ROMtype is array (integer range 2047 downto 0) of std_logic_vector(1 downto 0);
signal ROM      : ROMtype;

attribute syn_ramstyle of ROM : signal is "block_ram";

signal Addrtemp : std_logic_vector(10 downto 0);

begin
process(clk)
        begin
        if clk'event and clk = '1' then
                if write = '1' then
                        ROM(Conv_integer(Addr)) <= DataIn;
                end if;
                Addrtemp <= Addr;
        end if;
end process;

DataOut <= ROM(Conv_integer(Addrtemp));

end smallmem_synth;
```

Figure 4-16. VHDL Code for General ROM Component

4.4.5. Implementation Results

The generic AES architecture described has been captured using VHDL and numerous designs instantiated using Virtex-E XCV600E-8BG432 FPGAs. Xilinx Foundation Series 3.1i and Synplify Pro V6.0 software were used in the synthesis of these circuits. The results obtained, show that the 128-bit key implementation can operate at up to 310 Mbits/sec utilising 4681 CLB slices and 20 BRAMs. The 192-bit key design can run at a data-rate of 277 Mbits/sec and requires 4382 CLB slices and, again, 20 BRAMs. The 256-bit design is the largest of the three implementations, with the corresponding figures being 268 Mbits/sec, 4992 CLB slices and 20 BRAMs.

Interestingly, the 192-bit design is smaller than the 128-bit implementation. This is a consequence of the key expansion procedure. For a 128-bit key, every fourth word is passed through a LUT, with this occurring ten times. However, in the case of the 192-bit key there are only six instances of a word being passed through an LUT, resulting in a smaller design. Since each core utilises full 128-bit feedback, the speed of encryption when in OFB, CBC and CFB modes is identical to that obtained for the ECB mode. Therefore the results outlined above apply for all four supported modes of operation.

```
library IEEE;
use IEEE.std_logic_1164.all;
library unisim;
use unisim.vcomponents.all;

entity smallmem is
           port(    clk,write,reset : in std_logic;
                    Addr            : in std_logic_vector(10 downto 0);
                    DataIn          : in std_logic_vector(1 downto 0);
                    DataOut         : out std_logic_vector(1 downto 0));
end smallmem;

architecture smallmem_synth of smallmem is

component RAMB4_S2

port (    WE,EN,RST,CLK   : in std_logic;
          ADDR            : in std_logic_vector(10 downto 0);
          DI              : in std_logic_vector(1 downto 0);
          DO              : out std_logic_vector(1 downto 0));
end component;

signal Addrtemp : std_logic_vector(10 downto 0);

begin

SM : RAMB4_S2

port map(  WE => write,   EN => '1',    RST => reset,  CLK => clk,
           ADDR => Addr, DI => DataIn, DO => DataOut);

end smallmem_synth;
```

Figure 4-17. RAM Component Instantiated in VHDL

As discussed in chapter 3, previous implementations have been based on a 128-bit key, with the majority implemented using Xilinx Virtex devices. Table 4-2 summarises the main characteristics of these implementations and the characteristics of the generic architecture. From the table, it is observed that the 128-bit implementation presented here compares very favorably in terms of important parameters such as area and data rate. However, the key aspect of the approach is that it achieves this with much greater on-chip capability than previous implementations. This includes supporting:
– Both encryption and decryption
– Three feedback modes
– On-chip key scheduling

Table 4-2. Overall Comparison of 128-bit Key Rijndael FPGA Implementations

	ENC and DEC	3 KEYS	MODES	KEY SCHED. ON CHIP	DEVICE	AREA	SPEED 128-bit key Mbps
McLoone, McCanny	•	•	•	•	XCV600E	4681 slices 20 BRAMs	310
Mroczlowski [65]	-	-	-	•	ALTERA EPF10K250	1032 LCs 20 EABs	268
Chodowiec et al.[63]	•	-	-	-	XCV1000	2507 slices	414
Dandalis et al [23]	-	-	-	•	XCV1000	5673 slices	353
Elbirt et al. [17]	-	-	•	-	XCV1000	3528 slices	294

4.5. Conclusions

In this chapter a LUT-based design approach is described, whereby the complex and time-consuming operations of the Rijndael algorithm, i.e. the multiplicative inverse operations and multiplication and addition in $GF(2^8)$, are pre-computed for all possible input values and the results placed in LUTs. This methodology leads to very high-speed circuit implementations. The LUT-based fully pipelined Rijndael Virtex-E implementation presented in this chapter achieves a data-rate of 12 Gbits/sec, a factor 6 times faster than previous single-chip implementations on the XCV1000 device. The approach results in high area requirements and currently, there are no FPGA devices on which to implement the LUT-based decryptor design, which requires 370 BRAMs. However, Xilinx are in the process of introducing a new family of Virtex-II Pro FPGAs that will consist of devices with up to 556 BRAMs. Hence, it will be possible to implement the decryptor design on this technology. The LUT methodology is more suited to iterative Rijndael designs and both encryptor and decryptor iterative implementations are possible. The LUT-based encryption iterative Rijndael design, when implemented on the XCV400E device, proves a factor of 1.6 times faster than a typical design in which only the SubBytes transformation is mapped to a LUT. It is also faster than previous iterative implementations on XCV1000 devices.

Performance, size, cost functionality and design reusability must all be carefully considered in the design of encryption systems. Previous work, including the research outlined in chapter 3, on such hardware implementations has led to specific, one-off, solutions. This chapter extends this research by presenting a new generic architecture for implementing multi-functional Rijndael encryption cores in silicon. This allows the rapid creation of silicon solutions, which perform both encryption and decryption in ECB, CBC, CFB and OFB modes of operation. It also allows separate and dedicated cores for 128-bit, 192-bit and 256-bit key designs to be effortlessly synthesised. All these designs incorporate on-chip key scheduling and can be readily accommodated on state-of-the-art single chip FPGAs and PLDs. This has been demonstrated through implementation on a Xilinx Virtex-E FPGA since this device is well suited to implementations of the Rijndael algorithm. However, the architecture presented is also readily migratable to other FPGA/PLD technologies and indeed to ASIC implementation. Design studies, for example, indicate that the iterative 128-bit key design (which operates at 310 Mbits/sec on a Virtex-E device) will operate at Gbit/sec speeds when implemented on current ASIC technology. Thus, even higher performance implementations can be derived either in the form of standalone ASIC chips or embedded SoC solutions where performance and/or volumes demand this type of solution. Moreover, the FPGA implementations produced have a silicon area and throughput comparable with or better than previous one-off solutions.

All of the designs achieve throughputs that are more than sufficient for modern communications applications, such as current and next generation wireless products. The IEEE Wireless LAN Standard 802.11b requires the accommodation of transmission rates of just 11 Mbits/sec, whilst the more advanced 802.11g standard currently being developed requires data rates of up to 54 Mbits/sec.

In the future, Information Technology (IT) applications, such as wireless phones, wireless computing, pay-TV and audio/video copy protection schemes will be realised as embedded systems and therefore will rely heavily on security mechanisms [73]. Hence there is a real need for efficient, reusable multi-functional IP encryption cores such as those that can be easily derived from the generic AES architecture presented in this chapter.

Chapter 5

HASH ALGORITHMS AND SECURITY APPLICATIONS

5.1. Introduction

The security services required to guarantee a fully protected networking system are [76, 29]:
- *Confidentiality* - protecting the data from disclosure to unauthorised bodies
- *Authentication* - assuring that received data was indeed transmitted by the body identified as the source
- *Integrity* - maintaining data consistency and ensuring that data has not been altered by unauthorised persons
- *Non-repudiation* - preventing the originator of a message from denying transmission

Cryptographic mechanisms exist which provide these vital security services. Private-key and public-key encryption algorithms provide confidentiality and can provide authentication and integrity protection. Hash algorithms and digital signatures ensure authentication and integrity protection, while digital signatures also provide non-repudiation. Security applications and protocols typically implement one or more of these security mechanisms. One such application is the Internet Protocol Security (IPSec) standard.

The IPSec standard incorporates private-key and hash algorithms. In this chapter a novel single-chip hardware IPSec cryptographic design is presented, which comprises the generic Rijndael architecture described in chapter 4 and a HMAC-SHA-1 authentication algorithm design [77]. The IPSec core supports the cryptographic requirements of the IP Authentication

Header (AH) and Encapsulation Security Payload (ESP) and any combination of these two protocols. It is also capable of supporting any application requiring authentication and/or encryption, such as Wireless Local Area Networks (WLANs), the Secure Electronic Transaction (SET) and Transport Layer Security (TSL) protocols, Virtual Private Networks (VPNs) and firewalls.

The HMAC-SHA-1 design incorporates a low area and high speed SHA-1 architecture. This efficient architecture, which utilises a shift register design approach, is described in the chapter. Anticipating the increase in security which will be afforded by the AES, the NIST proposed an expansion of their hash standard, SHA-1, to include the SHA-256, SHA-384 and SHA-512 algorithms, which produce 256, 384 and 512-bit message digests respectively. The first hardware implementation of the SHA-384 and SHA-512 hash algorithms to be reported in the literature is also outlined [78].

Performance evaluations for the IPSec cryptographic core and SHA-384/SHA-512 design are presented. The integration of the IPSec core into other security applications and protocols is discussed. Finally, the chapter finishes with some important conclusions.

5.2. Internet Protocol Security (IPSec)

Internet protocols were first developed in the mid-1970s when the Defence Advanced Research Projects Agency (DARPA) became interested in establishing a packet-switched network that would facilitate communication between dissimilar computer systems [79]. DARPA funded research work by Stanford University and Bolt, Beranek and Newman (BBN), which resulted in the development of the Internet Protocol (IP) suite. The current version of this protocol is IP version 4 (IPv4). However, a newer version, IPv6, exists and is being utilised in limited portions of the Internet. IPv6 was designed to accommodate the increase in the Internet's popularity. Its address size is 128-bits whereas the IPv4 address size is 32-bits in length.

IPSec is an extension to the IP suite and is a globalised solution to the problem of Internet security. It is applied to the Network layer – layer 3 of the 7-layer Open Systems Interconnect (OSI) reference model. More specifically it operates on the Internet layer of the TCP/IP protocol suite and thus, provides inherent security to any application. Rather than requiring each email program or Web browser to implement its own security mechanism, IPSec involves a change to the underlying facilities that are used by each application [80] and also allows security to be applied to network traffic without involving end users. IPSec is described in several Request For Comments (RFCs) [81, 82, 83]. The protocol is designed to provide support for confidentiality, authentication and integrity. It also offers:

Access Control   -   restricts access of data to strictly authorised entities

Replay Protection   -   protects against replay attacks, in which a packet is extracted from the data stream and reused later by an attacker

The IPSec protocols include instructions for integration in both IPv4 and IPv6. However, in this chapter the protocols are discussed only in respect to IPv6. The two security mechanisms of IPSec are the Authentication Header (AH), which provides data origin authentication and connectionless integrity, and the Encapsulating Security Payload (ESP), which provides connectionless data confidentiality services [84]. Both the AH and ESP are used in accordance with a Security Association (SA). The SA is the agreement between two or more bodies on the security services they wish to utilise, such as which authentication algorithm, mode and keys to use in the AH mechanism and which encryption algorithm, mode and keys to use in the ESP mechanism. Therefore, an IP data packet can only be authenticated and/or decrypted if the receiver can correlate it to the appropriate SA.

IPSec supports two methods of operation, tunnel mode and transport mode. In transport mode, only the upper-layer protocol data segment of the IP packet, for example, a Transmission Control Protocol (TCP) segment, is authenticated or encrypted and it is typically used for end-to-end protection of data packets between two hosts. In tunnel mode the entire IP packet is authenticated or encrypted. The result is then transmitted within another IP packet which contains a new outer header. In effect, the entire original packet travels through a 'tunnel' from one point of an IP network to another. Tunnel mode can be used between firewalls to create a Virtual Private Network (VPN) [85].

Efficient key management is also an important aspect of IPSec. The default automated key management protocol chosen for employment with IPSec is the Internet Key Exchange (IKE) [86]. IKE is discussed in chapter 6.

5.2.1. IP Authentication Header

The authentication header provides support for data integrity and authentication of the IP packets [9]. The AH, illustrated in Figure 5-1, is described in Table 5-1 below.

Table 5-1. Description of Authentication Header

Field	Length	Function
Next Header	8-bit	Identifies the type of header that follows the AH header, e.g. ESP header or TCP header
Payload Length	8-bit	Specifies the length of the AH header in 32-bit words, minus 2
Reserved	16-bit	Set to 0, but reserved for future use
Security Parameter Index (SPI)	32-bit	Identifies the SA
Sequence Number	32-bit	Counter value which indicates number of messages sent from sender to receiver using current SA – allows replay protection
Authentication Data	variable	Contains the Message Authentication Code (MAC) for the data

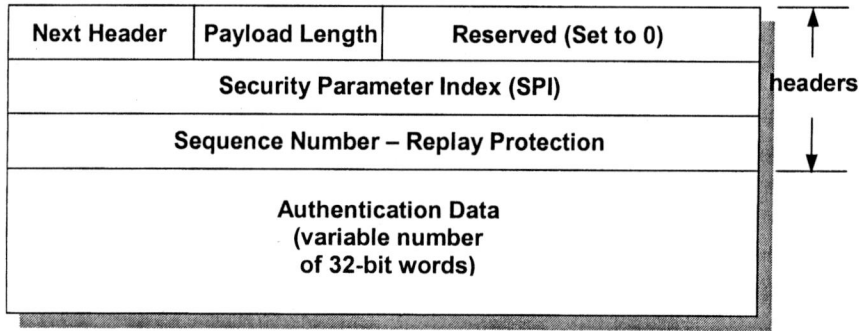

Figure 5-1. Outline of Authentication Header

The scope of authentication and the location of the AH varies between transport and tunnel mode. Figure 5-2 outlines a typical IPv6 header. Figure 5-3 shows the location of the authentication header and the authenticated fields for transport mode. The Message Authentication Code (MAC) is calculated over all the IP header fields which remain unaffected during transit or which are predictable in value by the receiving end-point. Header fields that will change in transit are set to zero for the MAC calculation, for example, the authentication data field. In transport mode, the AH is placed after the original IP header and associated extension headers and before the IP payload data, i.e. the TCP and packet data. The destination extension header can optionally be placed before or after the AH header. In tunnel mode the entire original IP packet and the new IP header and extension fields are authenticated except for those fields which change during transit. The AH header in tunnel mode is located between the new and original IP headers as illustrated in Figure 5-4.

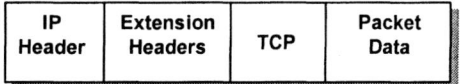

Figure 5-2. Typical IPv6 Header

Figure 5-3. AH Location in Transport Mode

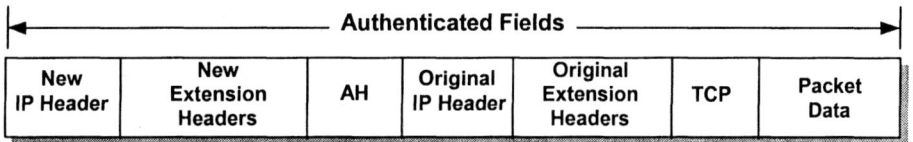

Figure 5-4. AH Location in Tunnel Mode

5.2.2. IP Encapsulating Security Payload

The IP AH does not transform the payload data of an IP packet and thus it remains unprotected to attacks such as eavesdropping. The Encapsulating Security Payload (ESP) mechanism offers data confidentiality, including message and limited traffic flow confidentiality [9]. ESP can also provide authentication if required by the user. The ESP header is outlined in Figure 5-5 and described in Table 5-2.

Similar to the AH, the location of the ESP header and scope of encryption and authentication varies for ESP in transport and tunnel mode. In transport mode, the payload data, padding, pad length and next header fields are encrypted. The ESP header is positioned after either the IP header or an AH header and ahead of the transport-layer header, as outlined in Figure 5-6. The padding, pad length and next header fields are collectively known as the *ESP trailer*, which is located after the IP data packet.

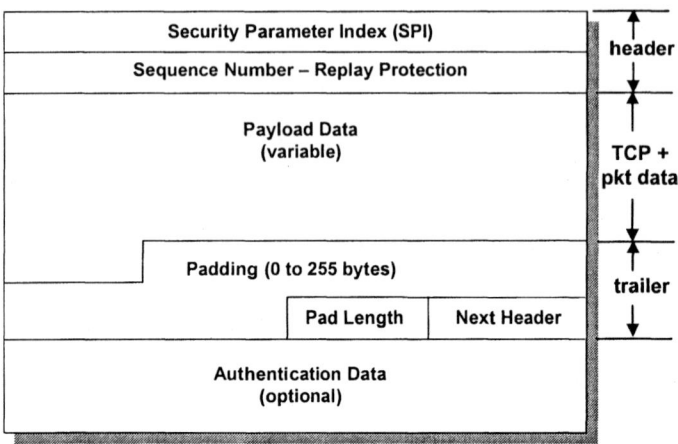

Figure 5-5. Outline of Encapsulating Security Payload

Table 5-2. Description of Encapsulating Security Payload

Field	Length	Function
Security Parameter Index (SPI)	32-bit	Identifies the SA
Sequence Number	32-bit	Counter value which indicates number of messages sent from sender to receiver using current SA – allows replay protection
Payload Data	variable	Refers to the data to be encrypted by the encryption algorithm specified by the SA. The data is a transport-level segment in transport mode and an entire IP packet in tunnel mode
Padding	0-255 bytes	Padding is used to ensure that the length of the data to be encrypted (payload data + pad length + next header) is an integral multiple of the encryption algorithm's input block size. Padding can also be used to disguise the message's true length and thus provide traffic flow confidentiality
Pad Length	8-bit	Specifies the length of the padding field
Next Header	8-bit	Identifies the type of header that follows the ESP header, e.g. TCP header or Extension header in IPv6
Authentication Data	variable	Optional field which contains the Message Authentication Code (MAC) for the data

Also if authentication is incorporated, only the IP payload is authenticated and is positioned after the ESP trailer. In tunnel mode the entire original IP packet is encrypted and optionally authenticated. The ESP header, in tunnel mode, is located between the new and original IP headers as illustrated in Figure 5-7.

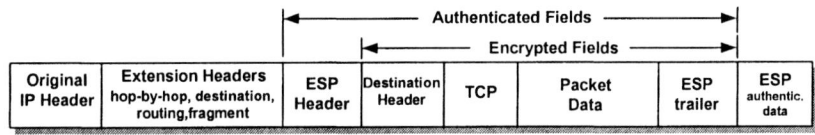

Figure 5-6. ESP Location in Transport Mode

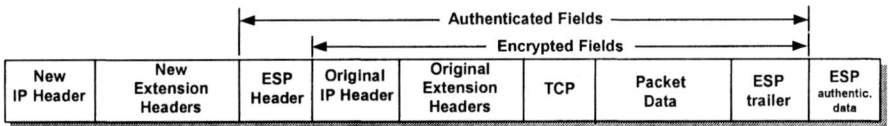

Figure 5-7. ESP Location in Tunnel Mode

5.3. IPSec Authentication Algorithms

The authentication algorithm employed by AH or ESP is indicated by the SA. The two authentication algorithms specified in the IP Authentication Header and Encapsulating Security Payload RFCs are the HMAC-MD5 and HMAC-SHA-1 algorithms. The SHA-1 and HMAC algorithms are described in sections 5.3.1 and 5.3.2 respectively.

5.3.1. SHA-1

The Secure Hash Algorithm (SHA-1) operates on a message in 512-bit blocks and produces a 160-bit message digest. The maximum message length acceptable is 2^{64} bits. SHA-1 is performed as follows:
- Pad the message to length $\equiv$ 448 mod 512 – padding is carried out by appending a single 1-bit followed by the required number of 0-bits
- Append the message length as a 64-bit binary number
- Parse message into N x 512-bit blocks of data
- Initialise 5 x 32-bit words, A, B, C, D and E such that,

```
A = 67452301
B = efcdab89
C = 98badcfe
D = 10325476
E = c3d2e1f0
```
(5.1)

- Perform 80 iterations of the SHA-1 processing function, outlined in Figure 5-8, on the first 512-bit data block
- The resulting 160-bit output initialises A, B, C, D and E for the processing function of the next data block
- After all *N* data blocks have been processed, the final output forms the 160-bit message digest

In the SHA-1 processing function, K_t are 32-bit constants where,

```
K_t = 5a827999     0 ≤ t ≤ 19
K_t = 6ed9eba1    20 ≤ t ≤ 39
K_t = 8f1bbcdc    40 ≤ t ≤ 59
K_t = ca62c1d6    60 ≤ t ≤ 79
```
(5.2)

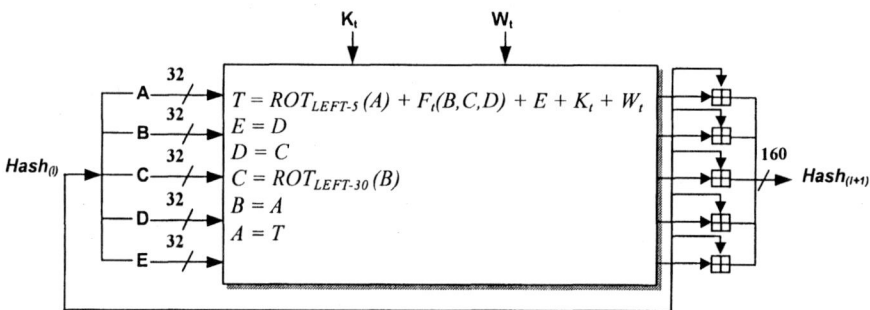

Figure 5-8. SHA-1 Processing Function

The message schedule, W_t, consists of 32-bit values such that,

$$W_t = \begin{cases} Message_t & 0 \leq t \leq 15 \\ ROT_{LEFT-1}(W_{t-3} \oplus W_{t-8} \oplus W_{t-14} \oplus W_{t-16}) & 16 \leq t \leq 79 \end{cases}$$ (5.3)

where ROT_{LEFT-n} (word) is a circular rotation of a word by n positions to the left.

The logical function used is,

$$F_t(x,y,z) = \begin{cases} (x \text{ AND } y) \text{ OR } (\overline{x} \text{ AND } z) & 0 \leq t \leq 19 \\ x \oplus y \oplus z & 20 \leq t \leq 39 \\ (x \text{ AND } y) \text{ OR } (x \text{ AND } z) \text{ OR } (y \text{ AND } z) & 40 \leq t \leq 59 \\ x \oplus y \oplus z & 60 \leq t \leq 79 \end{cases} \quad (5.4)$$

5.3.2. HMAC

The Keyed-Hash Message Authentication Code (HMAC) [87] authenticates both the source of a message and its integrity. The sender of a message uses the HMAC algorithm, which incorporates a secret key, to condense the message and produce a specific code. The message along with this code is sent to the recipient, who uses the same secret key and HMAC algorithm to recalculate the code. If the two codes are the same, the recipient is guaranteed that the message and its author are genuine.

The HMAC of a block of data can be calculated by performing the following equation:

$$\text{HMAC}(data) = H[k_o \oplus opad \mid\mid H[(k_o \oplus ipad) \mid\mid data]] \quad (5.5)$$

where,
H represents the Hash function, i.e. SHA-1
k_0 is the input key padded with zeros
$ipad$ = 0x36 (hexadecimal) x 64
$opad$ = 0x5a (hexadecimal) x 64

A 20-byte length key is mandated for HMAC-SHA-1 [88]. Since the HMAC-SHA-1 algorithm requires a key length of 64 bytes, the input key must be appended with 44 zero bytes. The HMAC of 160-bits is truncated to 96-bits for use with AH or ESP.

5.4. IPSec Cryptographic Processor Design

Many systems require both authentication and encryption protection. The obvious method of achieving this is to employ the ESP protocol with authentication since an individual SA can only implement either the AH or ESP protocols. Multiple SAs, however, can be utilised in order to implement both protocols. In fact, the RFC describing the IPSec security architecture includes a number of cases outlining combinations of SAs which must be supported by compliant IPSec hosts and security gateways. The cases mentioned employ both authentication and encryption by combining the AH and ESP protocols. Hence, the aim of the research outlined in this chapter was to achieve a single-chip design capable of performing the cryptographic requirements of both the AH and ESP mechanisms.

The encryption and authentication algorithms chosen for the IPSec design are Rijndael and HMAC-SHA-1 respectively. The DES algorithm operating in CBC mode, is currently the only algorithm mandated for the IPSec ESP header, although Triple DES is also commonly used. However, as described in chapter 1, the Rijndael algorithm replaced DES as the FIPS encryption standard in November 2001. This modification has yet to be reflected in the IPSec standards but it is expected that the IPSec working group will declare Rijndael as a mandatory encryption algorithm for the ESP protocol [80, 89, 90]. Predicting that this change will take place in the near future, the design presented here utilises the new Rijndael algorithm in implementing IPSec. The encryption core used in the IPSec design is the generic Rijndael architecture described in chapter 4. From this encryption architecture, cores can be generated for all three key lengths outlined in the Rijndael algorithm specification. Each design performs encryption and on-the-fly decryption and supports ECB, OFB, CBC and CFB modes of operation.

The two authentication algorithms specified for use in AH and ESP are HMAC-SHA-1 and HMAC-MD5. The HMAC-SHA-1 algorithm is chosen for implementation since the SHA-1 message digest is 32-bits longer than the MD5 digest and is considered stronger against brute force attacks [80, 9, 29]. In 1996 the MD5 algorithm was shown to be vulnerable to specific collision search attacks [91].

5.4.1. SHA-1 Design

The SHA-1 design consists of four main components - a padding block, message scheduler, compression block and controller - as illustrated in Figure 5-9.

The padding block is responsible for generating the padded 512-bit blocks required by the algorithm from the input data. The data is loaded in 32-bit blocks. A counter, with the aid of the LAST_BLK and LAST_DATA_CNT signals, is used to calculate the length of the entire message. If the message length is 448 bits an extra iteration of the compression block is performed in which the input comprises all zeros concatenated with a binary representation of the message length.

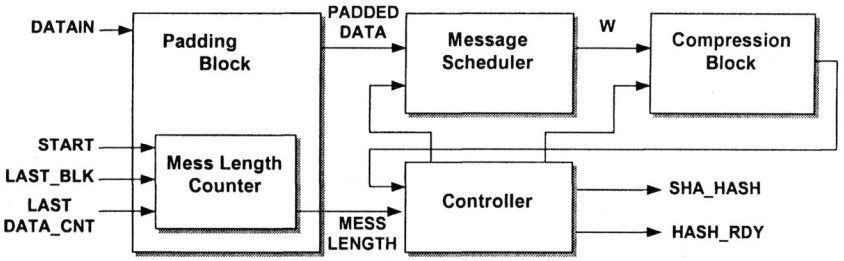

Figure 5-9. Outline of SHA-1 Algorithm Design

The message scheduler is implemented using a 16-stage shift register design. The use of shift registers in the implementation is motivated by the diagrammatical representation of hash algorithms provided by the NIST [92]. Figure 5-10 outlines the message scheduler design.

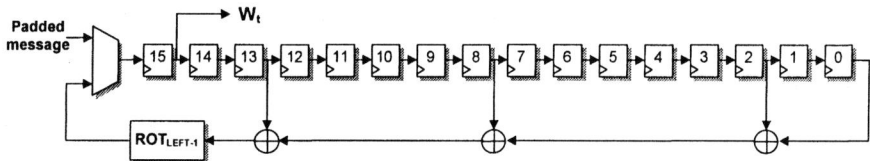

Figure 5-10. SHA-1 Message Scheduler Shift Register Design

For 16 clock cycles the registers are loaded with the 32-bit padded message blocks. On the next clock cycle the value of register 15 is replaced with the value resulting from equation (5.3) outlined in section 5.3.1. Since the output, W_t, is taken from between registers 14 and 15, and not from the output of register 0, an initial delay of 16 clock cycles is avoided.

The shift register methodology can also be utilised in the compression block design which implements the SHA-1 processing function, as depicted in Figure 5-11. The design uses 5 registers to store the continually updating values of A, B, C, D and E. The values of registers B, C and D are passed through one of four different functions every 20 cycles as outlined in equation (5.4).

The controller is responsible for controlling the flow of data throughout the overall design. It updates the values of A, B, C, D and E for each new compression iteration and carries out the addition between the *Hash*(i) values and the updated 32-bit words A to E, to form the new *Hash*(i+1). When the final compression is performed, the results of this addition are concatenated to form the 160-bit output hash.

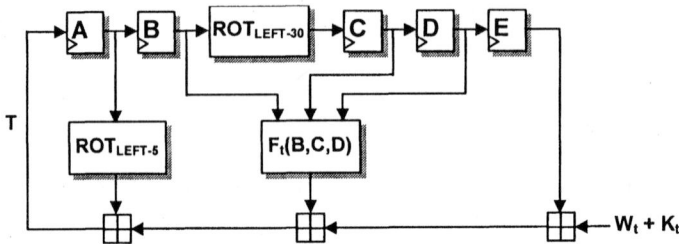

Figure 5-11. SHA-1 Compression Block Design

The MD5 and SHA-1 hash algorithms have many similar characteristics and therefore, this hardware design could easily be extended to incorporate MD5 functionality.

5.4.2. HMAC-SHA-1 Design

The HMAC-SHA-1 algorithm design incorporates the SHA-1 component as shown in Figure 5-12. The key-padding block extends the mandatory 20-byte input key to a 64-byte key as required by the HMAC algorithm. The padded key is then XORed with the constant *opad* and *ipad* 64-byte values. The concatenate blocks serve to append the XOR results, in the first instance with the DATAIN and in the second instance with the resulting 160-bit hash, to form the input to the SHA-1 hash function. Therefore, the HMAC algorithm utilises two passes of the underlying hash function, where the inputs are the ($k_0 \oplus ipad$) 512-bit block and the input data to achieve the first hash output. The ($k_0 \oplus opad$) result and the first hash output are the inputs which achieve the final hash or HMAC output. In order to maintain a low area design, and since the first hash output forms a part of the second hash function input, only one SHA-1 component is used with a multiplexor to select between the two inputs. The HMAC controller manages the flow of data throughout the circuit. Since it is necessary to wait until the hash function is performed twice, START and WAIT signals are used in order to control the input data. After the second SHA-1 iteration, the resulting 160-bit hash is truncated to obtain the final 96-bit HMAC output.

Alternatively, if the input key is not changed frequently, a more efficient implementation is possible. The intermediate results obtained by passing the ($k_0 \oplus ipad$) and ($k_0 \oplus opad$) blocks through the hash function can be pre-computed and used to initialise SHA-1's A, B, C, D and E values for the data input and first hash output blocks respectively. This will significantly reduce the computational time of producing the HMAC of a message. The design is especially efficient if the messages for which the MAC is computed are short [9]. The intermediate results can be stored using two 160-bit registers. However, this will restrict the design to authentication utilising just one key. If the key is altered, the design will need to be re-synthesised. A better approach is to input the intermediate results as an initialisation value, IVKEY, for the SHA-1 function. The efficient implementation of the HMAC-SHA-1 algorithm consists of the controller and SHA-1 component as depicted in Figure 5-13.

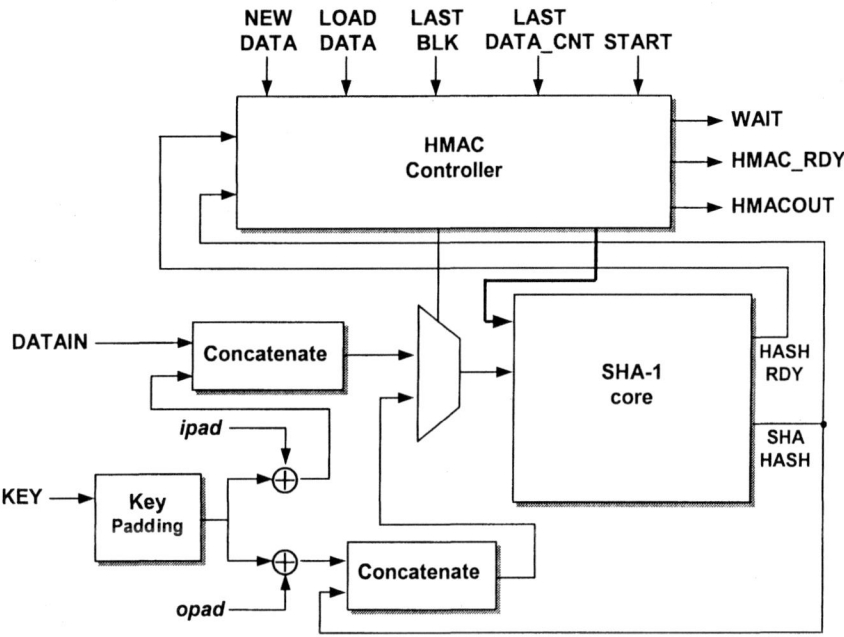

Figure 5-12. Outline of HMAC-SHA-1 Algorithm Design

5.4.3. Overall IPSec Design

An outline of the IPSec cryptographic core is given in Figure 5-14. The core comprises the Rijndael design and can include either of the two HMAC-SHA-1 designs. The ordinary HMAC core can be employed when frequent key changes are required whereas the more efficient HMAC design is suitable when the same key is used in any one data transfer session. The Rijndael and HMAC-SHA-1 designs are low-area designs to allow for the implementation of both on a single-chip. Hence, the overall IPSec cryptographic core can perform encryption and/or authentication. The scenario whereby authentication is performed on the encrypted data, the ciphertext, is also supported. This occurs when operating the ESP protocol with authentication. If the A_ESP_OPTION signal is high, ciphertext can be input into the HMAC-SHA-1 component.

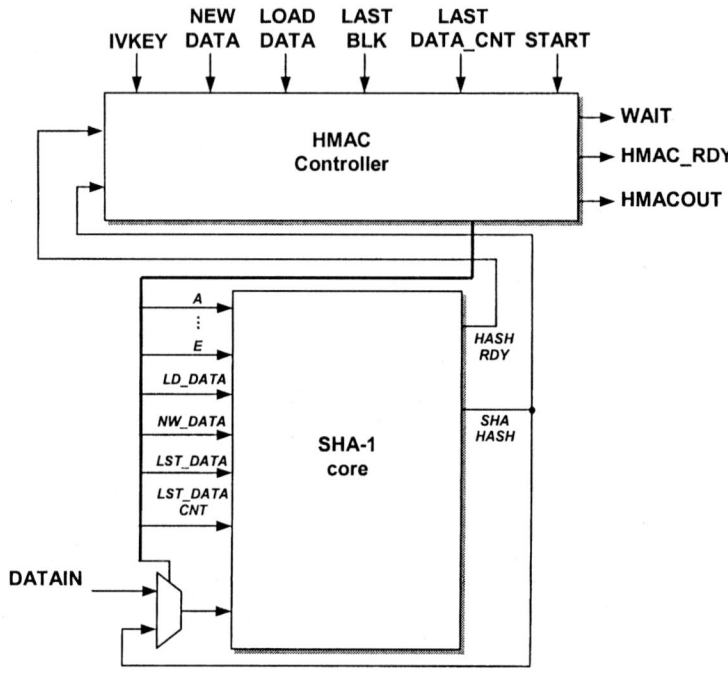

Figure 5-13. Outline of Efficient HMAC-SHA-1 Algorithm Design

5.5. Performance Results

The IPSec cryptographic design described was simulated using Modelsim XE and synthesised using Synplify Pro V6.0 and Xilinx Foundation Series 3.1i software. The SHA-1, HMAC-SHA-1 and Rijndael algorithm designs were individually verified using the test vectors provided in the FIPS 180-1 Secure Hash Standard [93], FIPS 198 HMAC Standard [94] and the Rijndael specification [56] respectively. The overall IPSec design was implemented on a single XCV1000E Xilinx Virtex-E device. 310 of 404 IOBs, 6143 CLB slices and 20 BRAMs are utilised. The BRAMs are required by the Rijndael design. The IPSec design is capable of performing 128-bit encryption and decryption at a rate of 310 Mbits/sec (clock speed – 24.2 MHz) as described in §4.4.5.

Hash Algorithms & Security Applications

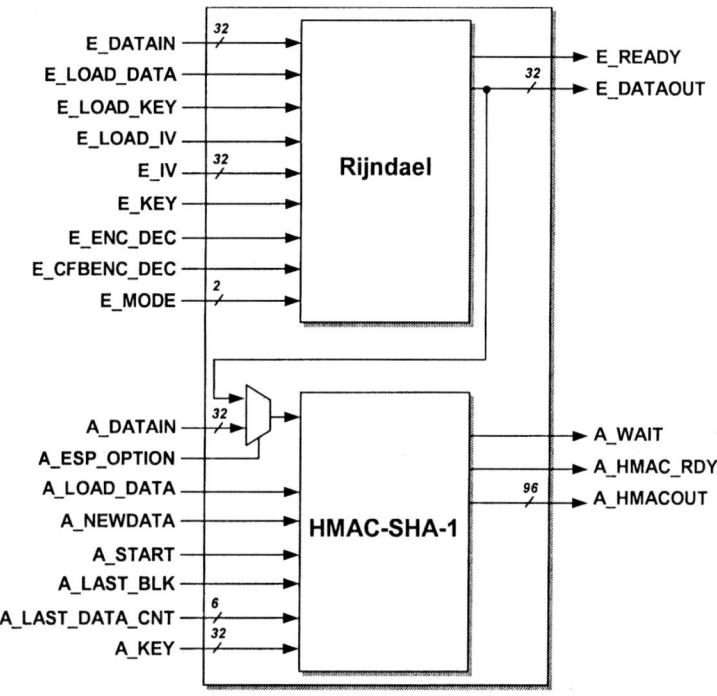

Figure 5-14. Outline of IPSec Cryptographic Core

It is difficult to obtain a true data throughput figure for the HMAC-SHA-1 element of the design since it creates a hash of an entire message of size $< 2^{64}$. Also, the SHA-1 design processing time is 82 clock-cycles, but if the final 512-bit padded block being processed contains >448 bits of the actual message, an extra iteration of 82 clock cycles is required. However, maximum and minimum data-rate figures are attainable. If the extra iteration is required, the HMAC-SHA-1 processing time is 410 clock cycles, i.e. five 512-bit blocks of data are passed through the SHA-1 function. The HMAC design can run at a clock speed of 58 MHz and therefore, will perform at a minimum data rate of 72 Mbits/sec. The typical HMAC-SHA-1 processing time is 328 clock cycles and hence the maximum data throughput is 90 Mbits/sec. In the efficient HMAC design, the maximum throughput is 199 Mbits/sec – only two passes of the hash function are required since the XORed key hash results are pre-computed, which implies a processing time of 163 clock cycles. The minimum data rate with which the efficient design can run is 133 Mbits/sec.

If both encryption and authentication are being performed at the same time, two separate clocks can be utilised with a FIFO to control the flow of data. However, if only one clock is used, the HMAC-SHA-1 design runs at

24.2 MHz and hence a data rate of 30 to 37.8 Mbits/sec. However, the efficient design can operate at 50 to 76 Mbits/sec.

Thus, both architectures are more than sufficient in providing security for 56 Kbit/sec (phone line modems), 1.54 Mbit/sec (T1 wireless), and 10 Mbit/sec (Ethernet) networks. The efficient HMAC design can also provide the security requirements of 100 Mbit/sec (Ethernet) networks. No other IPSec cryptographic hardware implementations comprising the Rijndael and HMAC-SHA-1 algorithms have been reported in the literature.

The HMAC-SHA-1 component incorporates a low area, yet very efficient SHA-1 algorithm implementation. The SHA-1 stand-alone architecture utilises just 1125 CLB slices and operates at 393 Mbits/sec with a clock rate of 63 MHz. For comparison purposes a number of commercial SHA-1 FPGA implementations are outlined in Table 5-3. The SHA-1 design described in this chapter is the smallest of these designs and is also the most efficient. The use of the 16-stage shift register in designing the message schedule and compression components leads to this highly efficient, compact implementation.

The IPSec cryptographic hardware design presented can also compliment the IP Payload Compression Protocol (IPComp) [95] in speeding up both the encryption and authentication involved in the AH and ESP protocols. IPComp employs data compression algorithms to reduce the IP packet payload size and hence packet transmission time. Therefore, since the actual amount of data being encrypted or authenticated is reduced, the execution times of the encryption and authentication algorithms will also be decreased [96].

Table 5-3. Comparison of SHA-1 algorithm FPGA implementations

Manufacturer	Device Used	CLB Slices	Data Rate Mbits/sec	Efficiency Slices/Data Rate
Helion [97]	VirtexE-8	1550	512	0.33
Alma [98]	VirtexE-8	1310	385	0.29
McLoone, McCanny [77]	VirtexE-8	1125	393	0.35

5.6. IPSec Cryptographic Processor Use in Other Applications

The IPSec design is also capable of supporting any application requiring both authentication and encryption. For example, it can be utilised to provide the security needs of Wireless LANs (WLANs), the Secure Electronic Transaction (SET) protocol, the Transport Layer Security (TLS) and the Wireless Transport Layer Security (WTLS) protocols.

The SET protocol is a mechanism for providing secure credit card payments over the Internet [99]. It was developed by Mastercard and Visa (with assistance from IBM, Microsoft, VeriSign and other companies) and published in 1996. The service is tailored to the specific needs of a particular application and thus is an application layer security protocol. SET uses DES encryption to provide confidentiality of information. For example, credit card numbers are issued to the bank but cannot be viewed by the retailer. Similarly to IPSec, it is inevitable that Rijndael will replace DES as the symmetric encryption algorithm in the SET protocol. The HMAC-SHA-1 algorithm is used in SET to authenticate the payment information, such as personal details and payment instructions, sent by a customer to the retailer.

The Secure Socket Layer (SSL) protocol was developed by Netscape to provide encrypted and authenticated communication between clients and servers. TLS is a standard proposed by the Internet Engineering Task Force (IETF) in an effort to produce an Internet standard version of SSL [9] and thus it is very similar to SSL version 3.0. It is a transport layer security protocol and runs above TCP/IP and below higher-level protocols such as the HyperText Transport Protocol (HTTP). The TLS protocol allows the authentication of a server and a client to each other and allows both machines to establish a secure encrypted connection. It includes two sub-protocols – the TLS record protocol and the TLS handshake protocol [100]. The handshake protocol negotiates the security parameters for the record layer and the key exchange. The record protocol provides connection security. This involves fragmentation of the upper-layer message. Compression is optionally applied to the fragmented blocks. The MAC is computed over the compressed data as follows:

HMAC (data) = $H[k_o \oplus$ opad $\| H[(k_o \oplus$ ipad) $\|$ seq_num $\|$
TLScompressed.type $\|$ TLScompressed.version $\|$
TLScompressed.length $\|$ TLScompressed.fragment]]   (5.6)

where,
    H represents the Hash function, i.e. SHA-1
    k_0 is the input key padded with zeros

ipad = 0x36 (hexadecimal) x 64
opad = 0x5a (hexadecimal) x 64
seq_num is the sequence number for the data block
TLScompressed.type is the higher-level protocol type
TLScompressed.version refers to the protocol version
TLScompressed.length is the length of the compressed fragment
TLScompressed.fragment represents the compressed/plaintext fragment

The compressed message and the HMAC result are encrypted using a symmetric encryption algorithm. A range of symmetric algorithms is supported by TLS such as DES, IDEA and Triple DES. In January 2002, an Internet draft was published by the IETF to incorporate the AES algorithm into TLS. The draft indicated that since the AES has withstood extensive cryptanalysis and proven to be efficient, it is in fact a very desirable choice of symmetric algorithm [101]. The AES algorithm in CBC mode is required.

The Wireless Transport Layer Security (WTLS) protocol was developed by the Wireless Application Protocol (WAP) forum in 2001 to provide security for applications that operated over wireless networks [102]. It is based on the TLS protocol and thus authentication and encryption are performed similarly to that previously described for TLS.

In recent years, there has been a tremendous growth in the wireless communications market. The IEEE 802.11 protocol, which specifies how to achieve wireless connectivity for fixed, portable and moving stations in a local area, is the principal standard for WLANs. It introduced the Wired Equivalent Privacy (WEP) protocol to protect wireless communication from eavesdropping and other attacks. However, WEP has been shown to be unsatisfactory [103, 104, 105]. WEP uses a cyclic redundancy checksum (CRC-32) to produce an integrity checksum for authentication and the RC4 algorithm for encryption. Borisov *et al.* [104] demonstrated the importance of using a cryptographic secure MAC such as HMAC-SHA-1 to provide authentication rather than using the insecure CRC checksum. Also, recognising the insecurities of WEP, the IEEE Task Group E has approved a draft Enhanced Security Network (ESN) protocol which uses AES over the weaker RC4 algorithm. It has also been suggested that IPSec could be employed to provide the security requirements for the IEEE 802.11 standard [105]. Indeed, some companies have already begun to utilise IPSec in wireless LAN systems [106].

The IPSec protocol itself is used to provide security in VPNs and firewalls. Therefore, the novel IPSec cryptographic design presented in this chapter, which comprises the Rijndael algorithm and HMAC-SHA-1 algorithm, can be used to provide numerous applications and protocols with strong, efficient authentication and encryption.

5.7. SHA-384/SHA-512 Processor

In anticipation of the expected increase in use of the AES standard, the NIST proposed the addition of three new hash algorithms to the Secure Hash Standard – SHA-256, SHA-384 and SHA-512 [107]. In this section SHA-384 and SHA-512 are discussed.

5.7.1. SHA-512

SHA-512 operates on a message in 1024-bit blocks and produces a 512-bit message digest. The maximum message length acceptable by the algorithm is 2^{128} bits. Similar to SHA-1, the SHA-512 algorithm consists of message padding and parsing, a message schedule and a processing function and it is carried out as follows:
- Pad the message to length $\equiv 896 \bmod 1024$
- Append the message length as a 128-bit binary number
- Parse message into N x 1024-bit blocks of data
- Initialise 8 x 64-bit words, A, B, C, D, E, F, G and H such that,

```
A = 6a09e667 f3bcc908
B = bb67ae85 84caa73b
C = 3c6ef372 fe94f82b
D = a54ff53a 5f1d36f1
E = 510e527f ade682d1
F = 9b05688c 2b3e6c1f
G = 1f83d9ab fb41bd6b
H = 5be0cd19 137e2179
```
(5.7)

- Perform 80 iterations of the SHA-512 processing function, outlined in Figure 5-15, on the first 1024-bit data block
- The resulting 512-bit output initialises A, B, C, D, E, F, G and H for the processing function of the next data block
- After all N data blocks have been processed, the final output forms the 512-bit message digest

In the SHA-512 processing function, K_t are a sequence of eighty 64-bit constants as outlined in Appendix E.
The message schedule, W_t, consists of 64-bit values such that,

$$W_t = \begin{cases} Message_t & 0 \leq t \leq 15 \\ \sigma_1^{512}(W_{t-2}) + W_{t-7} + \sigma_0^{512}(W_{t-15}) + W_{t-16} & 16 \leq t \leq 79 \end{cases} \quad (5.8)$$

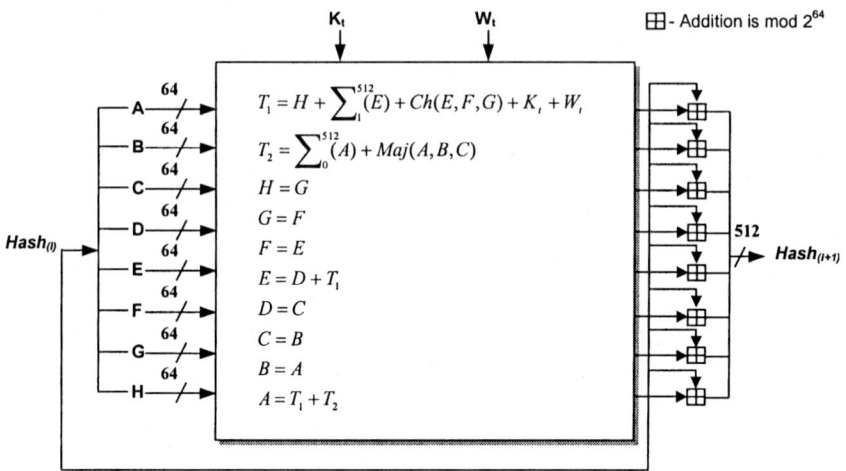

Figure 5-15. SHA-512 Processing Function

The logical functions used in the message schedule and processing function are,

$$Ch(x, y, z) = (x \text{ AND } y) \oplus (\bar{x} \text{ AND } z) \qquad (5.9)$$

$$Maj(x, y, z) = (x \text{ AND } y) \oplus (x \text{ AND } z) \oplus (y \text{ AND } z) \qquad (5.10)$$

$$\sum\nolimits_{0}^{512}(x) = ROT_{RIGHT-28}(x) \oplus ROT_{RIGHT-34}(x) \oplus ROT_{RIGHT-39}(x) \quad (5.11)$$

$$\sum\nolimits_{1}^{512}(x) = ROT_{RIGHT-14}(x) \oplus ROT_{RIGHT-18}(x) \oplus ROT_{RIGHT-41}(x) \quad (5.12)$$

$$\sigma_0^{512} = ROT_{RIGHT-1}(x) \oplus ROT_{RIGHT-8}(x) \oplus SHF_{RIGHT-7}(x) \qquad (5.13)$$

$$\sigma_1^{512} = ROT_{RIGHT-19}(x) \oplus ROT_{RIGHT-61}(x) \oplus SHF_{RIGHT-6}(x) \qquad (5.14)$$

where $ROT_{RIGHT-n}$ *(word)* is a circular rotation of a word by n positions to the right and $SHF_{RIGHT-n}$ *(word)* is the right shifting of a word by n positions.

5.7.2. SHA-384

The SHA-384 hash algorithm is almost identical to the SHA-512 algorithm. It also operates on 1024-bit blocks and the maximum message length acceptable is 2^{128} bits. The algorithms differ in initialisation values and in the message digest length. In SHA-384 the 8 x 64-bit words, A, B, C, D, E, F, G and H are initialised such that,

$$
\begin{aligned}
A &= \text{cbbb9d5d c1059ed8} \\
B &= \text{629a292a 367cd507} \\
C &= \text{9159015a 3070dd17} \\
D &= \text{152fecd8 f70e5939} \\
E &= \text{67332667 ffc00b31} \\
F &= \text{8eb44a87 68581511} \\
G &= \text{db0c2e0d 64f98fa7} \\
H &= \text{47b5481d befa4fa4}
\end{aligned}
\tag{5.15}
$$

The SHA-384 algorithm produces a 384-bit message digest which is formed by truncating the left-most 384 bits of the 512-bit hash output.

5.7.3. SHA-384/SHA-512 Design

Since the SHA-384 and SHA-512 algorithms are very similar they can both easily be implemented on a single-chip. Similar to SHA-1, the design includes a padding block, message scheduler, compression block and controller.

The controller controls the flow of data in the design. The padding block generates the padded 1024-bit data blocks required by the message scheduler. The entire message length is calculated as in the SHA-1 design using a counter and LAST_BLK and LST_DATA_CNT signals.

The shift register design approach is employed to implement the SHA-384/512 message scheduler and compression blocks. A 16-stage shift register architecture is used in the message scheduler design as shown in Figure 5-16.

The registers are loaded with the 64-bit padded message blocks over 16 clock cycles. Register 15 is then replaced with the result of equation (5.8) on the next clock cycle. This process continues for 80 clock cycles. Similar to the SHA-1 design, an initial 16 clock cycle delay is avoided by taking the output, W_t, from the output of register 15. The compression block architecture is depicted in Figure 5-17. Eight registers are used in the design to hold the values of A to H as they are updated on each cycle.

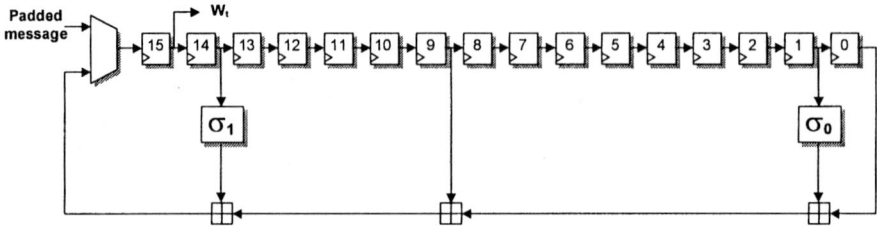

Figure 5-16. SHA-384/SHA-512 Message Scheduler Design

The functions $Ch(E, F, G)$, $Maj(A, B, C)$, $\Sigma_0(A)$ and $\Sigma_1(E)$ are as outlined in equations (5.9), (5.10), (5.11) and (5.12) respectively. The eighty 64-bit constants, K_t, are stored using a LUT. The LUT was implemented using two dual-port BRAMs since the target technology was a Virtex-E FPGA device. When the input data and write enable signals are set to zero, a BRAM can be used as a ROM or LUT. A counter, which counts to 80, is used to address the BRAM and the four 16-bit outputs are concatenated to form each 64-bit constant as it is required by the compression block. This BRAM implementation is outlined in Figure 5-18.

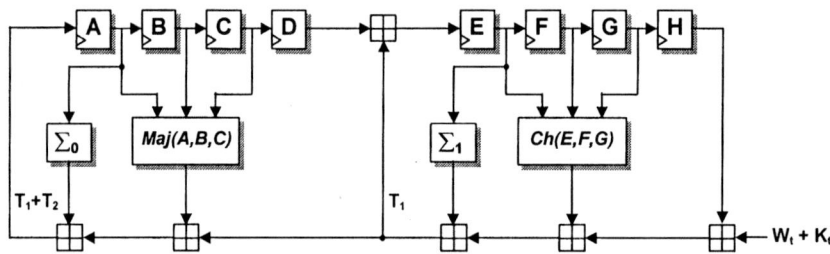

Figure 5-17. SHA-384/SHA-512 Compression Block Design

After 80 cycles the values in registers A to H are added to the initial hash values to obtain new hash values. These form the values of A to H for the next data block operation. When the final message block has been processed, the hash value outputs are concatenated to produce the 512-bit message digest.

To include SHA-384 capabilities, a multiplexor is added to select between the SHA-384 and SHA-512 initial values. A second multiplexor is utilised at the output to select between the 512-bit and the shortened 384-bit message digests.

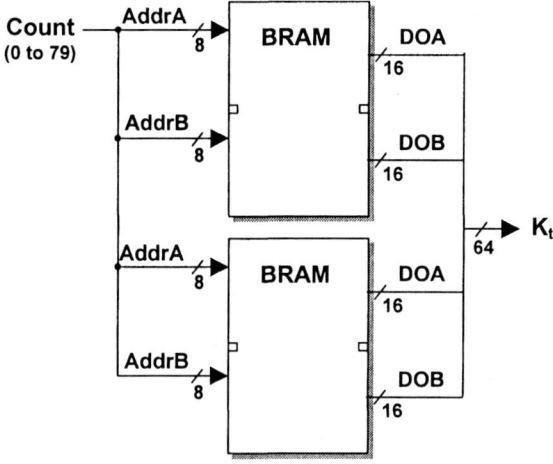

Figure 5-18. Implementation of Constants, Kt, using BRAMs

5.7.4. Performance Results

Hardware implementations of the SHA-256 algorithm exist [108, 109]. However, the design presented here is the first SHA-384/SHA-512 algorithm implementation. For implementation purposes the architecture described has been implemented on a Xilinx Virtex-E XCV600E-8 device. It was verified using the test vectors provided in the draft FIPS 180-2 standard [107]. The design requires 2914 CLB slices, 2 BRAMs and 141 IOBs. It operates at a clock speed of 38 MHz and thus has a throughput of 479 Mbits/sec. The shift register design approach and the use of BRAMs to store the eighty constants result in a compact and fast implementation. The message blocks are loaded in 64-bit blocks. Also the 384-bit or 512-bit hash result is output in 64-bit blocks. This leads to a lower IOB requirement and less routing and therefore improves the overall speed of the design.

5.8. Conclusions

In this chapter, a novel IPSec cryptographic chip design incorporating the Rijndael encryption algorithm and HMAC-SHA-1 authentication algorithm is presented. The design performs the encryption and authentication requirements of the IP AH and ESP protocols. The mandatory-to-implement IPSec encryption algorithm is the DES algorithm. However, it has recently been replaced in the FIPS standard by the AES algorithm, Rijndael. Therefore, it is reasonable to assume that it will also replace DES in the IPSec set of standards. Only one of the two mandatory authentication algorithms - the SHA-1 algorithm - was chosen for implementation in the

IPSec design since it is the stronger of the algorithms with a longer message digest. Also, it is believed that numerous algorithm options add to the complexity of the standard and inadvertently cause it to be less secure [89]. However, an extension to the design to include the MD5 algorithm can be easily achieved since the two algorithms share many similar computations.

Physical security and higher speeds are typical reasons for implementation on hardware devices such as FPGAs. The IPSec design is implemented using a single Virtex-E FPGA device for illustrative purposes of the possible speeds attainable. Encryption rates of 310 Mbits/sec and authentication at speeds of 90 Mbits/sec are achieved. If the efficient HMAC-SHA-1 design is included in the IPSec processor, authentication at a data-rate of 199 Mbits/sec is possible. However, the efficient design is only suitable when the same key is used in a data transfer session.

Therefore, the IPSec design is suitable for use in telephone line modem, T1 wireless and 10 Mbit/sec networks and if the efficient HMAC-SHA-1 design is utilised, the security requirements of 100 Mbit/sec networks can be supported. Also, if implemented on ASIC CMOS technology or if used in conjunction with the IPComp protocol, further improvements in overall performance are possible.

The IPSec design is also capable of providing the security requirements of many other applications and protocols. In particular, it can provide the cryptographic needs of the application layer security protocol, SET, and the transport layer protocols, TLS and WTLS. It can also provide security for the IEEE 108.11b WLAN standard. The IPSec protocol can be used in firewalls and is often considered the best VPN solution for IP environments as it includes strong security measures in its standards set [110].

The HMAC-SHA-1 design incorporates a highly efficient and compact SHA-1 architecture. The shift register design methodology used in the implementation of SHA-1's message scheduler and compression blocks helps to achieve the low area 393 Mbit/sec design.

The first reported SHA-384/SHA-512 algorithm single-chip hardware implementation is presented in the chapter. The design also employs shift registers in the implementation of the message scheduler and compression block. The message is input and the hash result is output in 64-bit blocks. Two BRAMs are used to store the eighty constants required by the compression block. This combined with the low IOB count and shift register design, leads to a high speed and low area implementation with a data rate of 479 Mbits/sec. Indeed if applications require encryption capabilities and authentication with longer message digests, the SHA-384/SHA-512 core can be integrated with the Rijndael design to achieve an efficient hardware implementation.

Overall, this chapter describes two new highly efficient single-chip hash algorithm implementations which provide different message digest lengths. It also presents an IPSec cryptographic system, which utilises a Rijndael architecture to provide encryption and a HMAC-SHA-1 architecture for authentication needs, and illustrates that these can readily be implemented using a single FPGA device.

Chapter 6
CONCLUDING SUMMARY AND FUTURE WORK

6.1. Concluding Summary

This book has examined in detail hardware architectures of the DES and Rijndael private key encryption algorithms. Numerous generic silicon architectures of both algorithms have been presented and comparisons provided with existing designs. Hash functions, which provide authentication, have been investigated and cryptographic applications that require both authentication and encryption were discussed. The main conclusions and contributions of each chapter are now summarised.

Chapter 1
The AES development effort, in which DES was replaced by the Rijndael algorithm in 2001, was described. The DES algorithm's 56-bit key was considered insufficient in providing adequate security for communication applications. The Rijndael algorithm was selected as the replacement to DES because it not only performed well in hardware and software implementations, but it also had low memory requirements, making it particularly suitable for area-restricted environments. In addition, it was easy to defend against timing and power analysis attacks and provided adequate security.

The advantages of implementing encryption algorithms in hardware and in particular using FPGAs, were presented. In comparison to software implementations, FPGA designs are naturally 10 to 100 times faster [111], can support parallel operations and provide additional physical security.

Although ASICs offer higher performance and are cheaper in high volumes, FPGAs have lower NRE costs, less fabrication delays and support in-circuit reprogrammability. Thus, FPGAs provide a much more efficient platform for cryptographic implementation than software devices and are rapidly becoming a very feasible alternative to ASICs.

Chapter 2

The DES symmetric key encryption algorithm, which has been in existence for over 20 years, was described in detail. DES is still required in many applications for backward compatibility purposes and will remain as a benchmark for newly developed algorithms in the future.

A new generic architecture has been described, from which DES designs can be created for use in numerous application requirements. These designs can provide ECB and CBC modes of operation, single or Triple-DES functionality, shifting and permutation key generation methods and a varying number of pipeline stages. Technology-independent, Virtex FPGA or Virtex-II FPGA implementation of the DES S-Boxes is also supported.

In addition, designs of specific speed and area configurations can be generated from the generic DES architecture by altering the number of pipeline stages. DES cores with 1, 2, 4, 8 and 16 pipeline stages have been compared and it has been shown that a fully pipelined, 16-stage design is the most efficient. The fully pipelined design implemented on a Virtex-E XCV300E device achieves a data-rate of 7.8 Gbits/sec. This is one of the fastest single-chip FPGA complete DES algorithm designs reported to date.

It has also been shown that the shifting DES key generation method is more efficient than the permutation method and is most suitable in designs requiring few key changes. The permutation method is more appropriate in designs involving frequent key changes as the shifting method incurs an initial latency on the input of a key during decryption.

A novel key scheduling approach has been presented that can be utilised in pipelined implementations of private-key encryption algorithms. DES is utilised to illustrate the key scheduling approach. The permutation key generation method is used to create the sub-keys, which are then delayed by an array of registers until they are required by each DES round. Therefore, the technique allows the loading of different keys every clock cycle.

Chapter 3

A new generic architecture has been described, which generates encryption-only Rijndael designs to support each of the three AES key lengths. An implementation of the 128-bit key Rijndael core operates at a rate of 7 Gbits/sec on a Virtex-E XCV812E device. At the time it was developed, this was the first fully pipelined Rijndael silicon design to be reported and it is among the fastest single-chip implementations reported to date.

The research undertaken also showed that Xilinx Virtex-E devices are particularly suitable for Rijndael algorithm implementation. The structure of a Virtex-E FPGA consists of adjacent columns of CLBs and BRAMS, which allow the algorithm components to be effectively placed resulting in very high throughputs.

An iterative Rijndael key schedule design has been described, which utilises only 2 BRAMs in implementation, regardless of the key length. The highly efficient design can be employed in both iterative and pipelined architectures.

A novel encryptor/decryptor Rijndael architecture has also been developed. It achieves a throughput of 3.2 Gbits/sec and is one of the first fully pipelined Rijndael FPGA implementations that can perform encryption and decryption. Excess memory utilisation is avoided by the addition of two initialising BRAMs – one is used to initialise all the BRAMs in the design with the required encryption values while the other provides the necessary decryption values.

Chapter 4

A LUT-based design methodology has been introduced whereby the complex finite field mathematical operations involved in the Rijndael algorithm are pre-computed for all possible inputs and the results stored in LUTs. A LUT-based 128-bit key fully pipelined architecture implemented on a Virtex-E XCV812E device boasts a pre-placement speed of 12 Gbits/sec. However, the technique incurs a high area penalty and as such, is more suitable for iterative implementations. The LUT-based 128-bit key iterative encryptor core runs at a throughput of 685 Mbits/sec, which is 1.6 times faster than the most competitive alternative implementation.

A hardware design for providing on-the-fly generation of the sub-keys required during Rijndael decryption has been demonstrated. This uses the final sub-key created during encryption as the inverse cipher key for decryption. Then, utilising an inverse key expansion procedure the decryption process sub-keys can be generated, as they are required by each inverse Round.

A new generic architecture for implementing multi-functional iterative Rijndael encryption cores in silicon has been presented. The generated cores support each of the three AES key lengths and can perform encryption and on-the-fly decryption in ECB, CBC, OFB and CFB modes of operation. The 128-bit key iterative core achieves a throughput of 310 Mbits/sec, which is comparable to existing iterative designs in terms of speed and area. Whereas previous Rijndael implementations have been one-off specific purpose designs, the advantage of this architecture is that it has much greater on-chip capability.

Chapter 5

A novel single-chip IPSec cryptographic design has been presented in this chapter. This comprises the multi-functional Rijndael architecture described in chapter 4 and a HMAC-SHA-1 authentication algorithm circuit. Currently, DES is the mandatory encryption algorithm in the IPSec standard, although it is expected that it will be replaced by Rijndael in the near future. There are two mandatory authentication algorithms, HMAC-SHA-1 and HMAC-MD5. However, SHA-1 has been shown to be the stronger and more secure of the two underlying hash functions and was therefore, selected for this work.

The HMAC-SHA-1 design authenticates information at a speed of 90 Mbits/sec. However, an efficient HMAC-SHA-1 design has been described, which achieves an authentication rate of 199 Mbits/sec. The efficient design assumes that the same key is used throughout a data transfer session.

An efficient, low area SHA-1 hardware design, which was utilised in the HMAC-SHA-1 architectures, has been presented. This is smaller than similar existing commercial SHA-1 implementations. The shift-register design approach used in both the message schedule and compression components leads to a highly compact implementation.

A shift register methodology was also used to develop one of the first hardware implementations of the SHA-384 and SHA-512 hash algorithms. Two BRAMs are utilised to store the algorithms' eighty constants and thus, a highly efficient, low area core with a data-rate of 479 Mbits/sec has been achieved.

6.2. Future work

6.2.1. Key Distribution

Key distribution is an important aspect in the provision of a secure cryptographic service. While symmetric key algorithms are typically used for bulk data encryption, asymmetric key algorithms are used in key distribution. It is vital to the security of a private key cryptosystem that the secret key is not compromised. Many key-exchange protocols exist which incorporate common public key algorithms such as RSA, ElGamal and Diffie-Hellman. Future work could involve a study of key-exchange protocols and an investigation into possible hardware implementations.

6.2.2. The Internet Key Exchange (IKE) Security Protocol

Internet Key Exchange (IKE) is a hybrid key management protocol, which is typically used with the IPSec standard. It is employed to generate, exchange and establish keys in a secure manner between two hosts of a network. It implements the Oakley and Skeme key exchange protocols in association with the Internet Security Association and Key Management Protocol (ISAKMP). The negotiation involves two phases. In phase 1, an ISAKMP security association (SA) is established, through which secure communication can take place between the two hosts. In phase 2, IPSec SAs are established - an inbound SA and an outbound SA are negotiated [80].

Oakley is a key exchange protocol based on the Diffie-Hellman algorithm, which provides services such as identity protection and authentication.

Skeme is a key exchange protocol that provides anonymity, repudiability and quick key refreshment [86].

ISAKMP is an abstract framework for key management protocols. It defines payload formats, provides specific protocol support and negotiates SAs.

Cryptographic algorithms, which are mandatory for use with IKE include:
- DES operating in CBC mode - it is probable that Rijndael will replace DES as the mandatory encryption algorithm
- Diffie-Hellman
- HMAC-SHA-1
- HMAC-MD5
- RSA

It is evident from the work presented in this book that highly efficient hardware implementations of DES, Rijndael and HMAC-SHA-1 are possible. Indeed, since the MD5 algorithm is similar to SHA-1, it is reasonable to assume that it too would perform well in hardware. The Diffie-Hellman and RSA public key algorithms contain complex modular exponentiation operations, which are resource intensive. It has been found that implementation of these operations in hardware greatly increases the speed of key exchange and significantly reduces latency [112]. Hardware implementations also support parallel processing, which would allow for the maintenance of a high number of simultaneous SAs. For these reasons, it would be interesting to develop a silicon architecture that would implement the IKE and provide support for the mandatory cryptographic algorithms.

6.2.3. Elliptic Curve Cryptography (ECC)

Elliptic curves were independently proposed in 1985 by Koblitz and Miller as viable public key cryptosystems [113]. Since their development, they are fast becoming direct competition for the RSA algorithm. They offer the highest strength-per-key-bit of any known public key system [114]. An ECC key, although 10 times smaller than an RSA key, will provide equivalent security. Hardware implementations of the RSA public key algorithm typically achieve poor throughputs. ECC architectures have already been implemented illustrating significantly better data-rates than that achieved by RSA [115].

ECCs are also proving to be an ideal choice in the emerging technology of smart cards, since they provide low area, low power, yet efficient implementations.

Further work could be performed on the design of generic silicon ECC architectures and the efficiency of these architectures in contrast to RSA designs could be examined.

Appendix A – Modulo Arithmetic

A.1. Modulo Division

The product of two polynomials $a(x) = a_3 x^3 + a_2 x^2 + a_1 x + a_0$ and $b(x) = b_3 x^3 + b_2 x^2 + b_1 x + b_0$ is $c(x)$ where,

$$c(x) = c_6 x^6 + c_5 x^5 + c_4 x^4 + c_3 x^3 + c_2 x^2 + c_1 x + c_0 \qquad (A.1)$$

The result, $c(x)$, is reduced modulo a polynomial of degree 4 by $M(x) = x^4 + 1$, as depicted in Figure A.1.

$$
\begin{array}{r}
c_6 x^2 + c_5 x + c_4 \\
x^4+1 \enclose{longdiv}{c_6 x^6 + c_5 x^5 + c_4 x^4 + c_3 x^3 + c_2 x^2 + c_1 x + c_0} \\
\underline{c_6 x^6 + c_6 x^2 } \\
c_5 x^5 + c_4 x^4 + c_3 x^3 + (c_2+c_6) x^2 + c_1 x + c_0 \\
\underline{c_5 x^5 + c_5 x } \\
c_4 x^4 + c_3 x^3 + (c_2+c_6) x^2 + (c_1+c_5) x + c_0 \\
\underline{c_4 x^4 + c_4} \\
c_3 x^3 + (c_2+c_6) x^2 + (c_1+c_5) x + (c_0+c_4)
\end{array}
$$

Figure A-1. Modulo Division by $M(x) = x^4 + 1$

The result of $c(x)$ modulo $x^4 + 1$ can be represented by,

$$d(x) = d_3 x^3 + d_2 x^2 + d_1 x + d_0 \qquad (A.2)$$

Therefore,

$$d_0 = c_0 \oplus c_4 \qquad d_2 = c_2 \oplus c_6$$
$$d_1 = c_1 \oplus c_5 \qquad d_3 = c_3$$

A.2. Multiplicative Inverse of a Byte

The inverse of the byte 1100 1011 (0xCB) is as follows:

Table A-1. Multiplicative Inverse of 0xCB in $GF(2^8)$

Row	Remainder	Quotient	Auxiliary
1	$m(x) = x^8 + x^4 + x^3 + x + 1$	-	0
2	$a(x) = x^7 + x^6 + x^3 + x + 1$	-	1
3	$x^6 + x^2 + x$	$x + 1$	$x + 1$
4	1	$x + 1$	x^2

The calculation $m(x)/a(x)$ is as shown in Figure A.2.

$$
\begin{array}{r}
x + 1 \\
x^7 + x^6 + x^3 + x + 1 \overline{\smash{)} x^8 + x^4 + x^3 + x + 1} \\
-(x^8 + x^7 + x^4 + x^2 + x) \\
\hline
x^7 + x^3 + x^2 + 1 \\
-(x^7 + x^6 + x^3 + x + 1) \\
\hline
x^6 + x^2 + x
\end{array}
$$

Figure A-2. Division of m(x)/a(x)

The resulting remainder, $x^6 + x^2 + x$ and the quotient, $x + 1$ are placed in the corresponding columns in row 3. This quotient is multiplied by the auxiliary value in row 2 and added to the auxiliary value in row 1. The result, $x + 1$ is placed in the auxiliary column of row 3. The polynomial, $a(x)$ is now divided by the remainder in row 3, as outlined in Figure A.3.

Once again, the remainder, 1 and the quotient $x + 1$ are placed in the corresponding columns in row 4. The auxiliary value of row 4 is, $(x + 1)(x + 1) + 1 = x^2 + 2x + 2$. The coefficients must lie in the finite field, {0,1}, hence, the auxiliary value is equal to x^2. Since the remainder is 1, x^2 is also the required inverse.

$$
\begin{array}{r}
x+1 \\
x^6+x^2+x \overline{\smash{\big)}\, x^7+x^6+\ x^3+x+1} \\
-(x^7+\ x^3+\ x^2\ ) \\
\hline
x^6+\ x^2+x+1 \\
-(x^6+\ x^2+x\ ) \\
\hline
1
\end{array}
$$

Figure A-3. Division of a(x) by remainder $x^6 + x^2 + x$

Appendix B – DES Algorithm Permutations and S-Boxes

B.1. Expansion and P-Permutations

The expansion and P-Permutation used in a DES round are given in Table B.1 and Table B.2 respectively.

Table B-1. Expansion Permutation

32	1	2	3	4	5	4	5	6	7	8	9	8	9	10	11
12	13	12	13	14	15	16	17	16	17	18	19	20	21	20	21
22	23	24	25	24	25	26	27	28	29	28	29	30	31	32	1

Table B-2. P-Permutation

16	7	20	21	29	12	28	17	1	15	23	26	5	18	31	10
2	8	24	14	32	27	3	9	19	13	30	6	22	11	4	25

B.2. S-Boxes

Tables B.3 to B.10 are the DES algorithm S-Boxes.

Table B-3. DES Algorithm S-Box 1

S-Box 1															
14	4	13	1	2	15	11	8	3	10	6	12	5	9	0	7
0	15	7	4	14	2	13	1	10	6	12	11	9	5	3	8
4	1	14	8	13	6	2	11	15	12	9	7	3	10	5	0
15	12	8	2	4	9	1	7	5	11	3	14	10	0	6	13

Table B-4. DES Algorithm S-Box 2

S-Box 2															
15	1	8	14	6	11	3	4	9	7	2	13	12	0	5	10
3	13	4	7	15	2	8	14	12	0	1	10	6	9	11	5
0	14	7	11	10	4	13	1	5	8	12	6	9	3	2	15
13	8	10	1	3	15	4	2	11	6	7	12	0	5	14	9

Table B-5. DES Algorithm S-Box 3

S-Box 3															
10	0	9	14	6	3	15	5	1	13	12	7	11	4	2	8
13	7	0	9	3	4	6	10	2	8	5	14	12	11	15	1
13	6	4	9	8	15	3	0	11	1	2	12	5	10	14	7
1	10	13	0	6	9	8	7	4	15	14	3	11	5	2	12

Table B-6. DES Algorithm S-Box 4

S-Box 4															
7	13	14	3	0	6	9	10	1	2	8	5	11	12	4	15
13	8	11	5	6	15	0	3	4	7	2	12	1	10	14	9
10	6	9	0	12	11	7	13	15	1	3	14	5	2	8	4
3	15	0	6	10	1	13	8	9	4	5	11	12	7	5	14

Table B-7. DES Algorithm S-Box 5

S-Box 5															
2	12	4	1	7	10	11	6	8	5	3	15	13	0	14	9
14	11	2	12	4	7	13	1	5	0	15	10	3	9	8	6
4	2	1	11	10	13	7	8	15	9	12	5	6	3	0	14
11	8	12	7	1	14	2	13	6	15	0	9	10	4	2	3

Table B-8. DES Algorithm S-Box 6

S-Box 6															
12	1	10	15	9	2	6	8	0	13	3	4	14	7	5	11
10	15	4	2	7	12	9	5	6	1	13	14	0	11	3	8
9	14	15	5	2	8	12	3	7	0	4	10	1	13	11	6
4	3	2	12	9	5	15	10	11	14	1	7	6	0	8	13

Table B-9. DES Algorithm S-Box 7

S-Box 7															
4	11	2	14	15	0	8	13	3	12	9	7	5	10	6	1
13	0	11	7	4	9	1	10	14	3	5	12	2	15	8	6
1	4	11	13	12	3	7	14	10	15	6	8	0	5	9	2
6	11	13	8	1	4	10	7	9	5	0	15	14	2	3	12

Table B-10. DES Algorithm S-Box 8

S-Box 8															
13	2	8	4	6	15	11	1	10	9	3	14	5	0	12	7
1	15	13	8	10	3	7	4	12	5	6	11	0	14	9	2
7	11	4	1	9	12	14	2	0	6	10	13	15	3	5	8
2	1	14	7	4	10	8	13	15	12	9	0	3	5	6	11

B.3. Key Scheduling Permutation to Remove Parity Bits

Table B.11 outlines the permutation utilised to remove the parity bits in the key scheduling process.

Table B-11. Key Permutation

57	49	41	33	25	17	9	1	58	50	42	34	26	18
10	2	59	51	43	35	27	19	11	3	60	52	44	36
63	55	17	39	31	23	15	7	62	54	46	38	30	22
14	6	61	53	45	37	29	21	13	5	28	20	12	4

Appendix C – LUTs Utilised in Rijndael Algorithm

C.1. Encryption LUT Values

The Hexadecimal values contained in the LUT utilised during Rijndael encryption are as outlined in the Table C.1. For example, an input of 0x00 would return the output, 0x63, an input of 0x08 would return the output, 0x30.

C.2. Decryption LUT Values

The Hexadecimal values contained in the Inverse LUT utilised during Rijndael decryption are as shown in the Table C.2. For example, an input of 0x00 would return the output, 0x52.

Table C-1. LUT Utilised During Rijndael Encryption

	0	1	2	3	4	5	6	7
0	63	7C	77	7B	F2	6B	6F	C5
1	CA	82	C9	7D	FA	59	47	F0
2	B7	FD	93	26	36	3F	F7	CC
3	04	C7	23	C3	18	96	05	9A
4	09	83	2C	1A	1B	6E	5A	A0
5	53	D1	00	ED	20	FC	B1	5B
6	D0	EF	AA	FB	43	4D	33	85
7	51	A3	40	8F	92	9D	38	F5
8	CD	0C	13	EC	5F	97	44	17
9	60	81	4F	DC	22	2A	90	88
A	E0	32	3A	0A	49	06	24	5C
B	E7	C8	37	6D	8D	D5	4E	A9
C	BA	78	25	2E	1C	A6	B4	C6
D	70	3E	B5	66	48	03	F6	0E
E	E1	F8	98	11	69	D9	8E	94
F	8C	A1	89	0D	BF	E6	42	68

	8	9	A	B	C	D	E	F
0	30	01	67	2B	FE	D7	AB	76
1	AD	D4	A2	AF	9C	A4	72	C0
2	34	A5	E5	F1	71	D8	31	15
3	07	12	80	E2	EB	27	B2	75
4	52	3B	D6	B3	29	E3	2F	84
5	6A	CB	BE	39	4A	4C	58	CF
6	45	F9	02	7F	50	3C	9F	A8
7	BC	B6	DA	21	10	FF	F3	D2
8	C4	A7	7E	3D	64	5D	19	73
9	46	EE	B8	14	DE	5E	0B	DB
A	C2	D3	AC	62	91	95	E4	79
B	6C	56	F4	EA	65	7A	AE	08
C	E8	DD	74	1F	4B	BD	8B	8A
D	61	35	57	B9	86	C1	1D	9E
E	9B	1E	87	E9	CE	55	28	DF
F	41	99	2D	0F	B0	54	BB	16

Table C-2. Inverse LUT Utilised During Decryption

	0	1	2	3	4	5	6	7
0	52	09	6A	D5	30	36	A5	38
1	7C	E3	39	82	9B	2F	FF	87
2	54	7B	94	32	A6	C2	23	3D
3	08	2E	A1	66	28	D9	24	B2
4	72	F8	F6	64	86	68	98	16
5	6C	70	48	50	FD	ED	B9	DA
6	90	D8	AB	00	8C	BC	D3	0A
7	D0	2C	1E	8F	CA	3F	0F	02
8	3A	91	11	41	4F	67	DC	EA
9	96	AC	74	22	E7	AD	35	85
A	47	F1	1A	71	1D	29	C5	89
B	FC	56	3E	4B	C6	D2	79	20
C	1F	D0	A8	33	88	07	C7	31
D	60	51	7F	A9	19	B5	4A	0D
E	A0	E0	3B	4D	AE	2A	F5	B0
F	17	2B	04	7E	BA	77	D6	26

	8	9	A	B	C	D	E	F
0	BF	40	A3	9E	81	F3	D7	FB
1	34	8E	43	44	C4	DE	E9	CB
2	EE	4C	95	0B	42	FA	C3	4E
3	76	5B	A2	49	6D	8B	D1	25
4	D4	A4	5C	CC	5D	65	B6	92
5	5E	15	46	57	A7	8D	9D	84
6	F7	E4	58	05	B8	B3	45	06
7	C1	AF	BD	03	01	13	8A	6B
8	97	F2	CF	CE	F0	B4	E6	73
9	E2	F9	37	E8	1C	75	DF	6E
A	6F	B7	62	0E	AA	18	BE	1B
B	9A	DB	C0	FE	78	CD	5A	F4
C	B1	12	10	59	27	80	EC	5F
D	2D	E5	7A	9F	93	C9	9C	EF
E	C8	EB	BB	3C	83	53	99	61
F	E1	69	14	63	55	21	0C	7D

Appendix D - LUTs in LUT-Based Rijndael Architecture

D.1. Encryption LUT Values

Table D.1 contains the values of the ByteSub LUT multiplied in $GF(2^8)$ by the hexadecimal number, 0x02 (*S[a]* • *02* or LUT_02).

Table D.2 contains the values of the ByteSub LUT multiplied in $GF(2^8)$ by the hexadecimal number, 0x03 (*S[a]* • *03* or LUT_03).

D.2. Decryption LUT Values

Table D.3 contains the values of the LUT_0D look-up table required in decryption, which is created by multiplying every possible byte from 0x00 to 0x11 by 0x0D in $GF(2^8)$.

Table D.4 contains the values of the LUT_09 look-up table required in decryption, which is created by multiplying every possible byte 0x00 to 0x11 by 0x09 in $GF(2^8)$.

Table D.5 contains the values of the LUT_0B look-up table required in decryption, which is created by multiplying every possible byte from 0x00 to 0x11 by 0x0B in $GF(2^8)$.

Table D.6 contains the values of the LUT_0E look-up table required in decryption, which is created by multiplying every possible byte from 0x00 to 0x11 by 0x0E in $GF(2^8)$.

Table D-1. LUT_02 Utilised During Encryption

	0	1	2	3	4	5	6	7
0	C6	F8	EE	F6	FF	D6	DE	91
1	8F	1F	89	FA	EF	B2	8E	FB
2	75	E1	3D	4C	6C	7E	F5	83
3	08	95	46	9D	30	37	0A	2F
4	12	1D	58	34	36	DC	B4	5B
5	A6	B9	00	C1	40	E3	79	B6
6	BB	C5	4F	ED	86	9A	66	11
7	A2	5D	80	05	3F	21	70	F1
8	81	18	26	C3	BE	35	88	2E
9	C0	19	9E	A3	44	54	3B	0B
A	DB	64	74	14	92	0C	48	B8
B	D5	8B	6E	DA	01	B1	9C	49
C	6F	F0	4A	5C	38	57	73	97
D	E0	7C	71	CC	90	06	F7	1C
E	D9	EB	2B	22	D2	A9	07	33
F	03	59	09	1A	65	D7	84	D0

	8	9	A	B	C	D	E	F
0	60	02	CE	56	E7	B5	4D	EC
1	41	B3	5F	45	23	53	E4	9B
2	68	51	D1	F9	E2	AB	62	2A
3	0E	24	1B	DF	CD	4E	7F	EA
4	A4	76	B7	7D	52	DD	5E	13
5	D4	8D	67	72	94	98	B0	85
6	8A	E9	04	FE	A0	78	25	4B
7	63	77	AF	42	20	E5	FD	BF
8	93	55	FC	7A	C8	BA	32	E6
9	8C	C7	6B	28	A7	BC	16	AD
A	9F	BD	43	C4	39	31	D3	F2
B	D8	AC	F3	CA	CF	F4	47	10
C	CB	A1	E8	3E	96	61	0D	0F
D	C2	6A	AE	69	17	99	3A	27
E	2D	3C	15	C9	87	AA	50	A5
F	82	29	5A	1E	7B	A8	6D	2C

Table D-2. LUT_03 Utilised During Encryption

	0	1	2	3	4	5	6	7
0	A5	84	99	8D	0D	BD	B1	54
1	45	9D	40	87	15	EB	C9	0B
2	C2	1C	AE	6A	5A	41	02	4F
3	0C	52	65	5E	28	A1	0F	B5
4	1B	9E	74	2E	2D	B2	EE	FB
5	F5	68	00	2C	60	1F	C8	ED
6	6B	2A	E5	16	C5	D7	55	94
7	F3	FE	C0	8A	AD	BC	48	04
8	4C	14	35	2F	E1	A2	CC	39
9	A0	98	D1	7F	66	7E	AB	83
A	3B	56	4E	1E	DB	0A	6C	E4
B	32	43	59	B7	8C	64	D2	E0
C	D5	88	6F	72	24	F1	C7	51
D	90	42	C4	AA	D8	05	01	12
E	38	13	B3	33	BB	70	89	A7
F	8F	F8	80	17	DA	31	C6	B8

	8	9	A	B	C	D	E	F
0	50	03	A9	7D	19	62	E6	9A
1	EC	67	FD	EA	BF	F7	96	5B
2	5C	F4	34	08	93	73	53	3F
3	09	36	9B	3D	26	69	CD	9F
4	F6	4D	61	CE	7B	3E	71	97
5	BE	46	D9	4B	DE	D4	E0	4A
6	CF	10	06	81	F0	44	BA	E3
7	DF	C1	75	63	30	1A	0E	6D
8	57	F2	82	47	AC	E7	2B	95
9	CA	29	D3	3C	79	E2	1D	76
A	5D	6E	EF	A6	A8	A4	37	8B
B	B4	FA	07	AF	25	8E	E9	18
C	23	7C	9C	21	DD	DC	86	85
D	A3	5F	F9	D0	91	58	27	B9
E	B6	22	92	20	49	FF	78	7A
F	C3	B0	77	11	CB	FC	D6	3A

Table D-3. LUT_0D Utilised During Decryption

	0	1	2	3	4	5	6	7
0	00	0D	1A	17	34	39	2E	23
1	D0	DD	CA	C7	E4	E9	FE	F3
2	BB	B6	A1	AC	8F	82	95	98
3	6B	66	71	7C	5F	52	45	48
4	6D	60	77	7A	59	54	43	4E
5	BD	B0	A7	AA	89	84	93	9E
6	D6	DB	CC	C1	E2	EF	F8	F5
7	06	0B	1C	11	32	3F	28	25
8	DA	D7	C0	CD	EE	E3	F4	F9
9	0A	07	10	1D	3E	33	24	29
A	61	6C	7B	76	55	58	4F	42
B	B1	BC	AB	A6	85	88	9F	92
C	B7	BA	AD	A0	83	8E	99	94
D	67	6A	7D	70	53	5E	49	44
E	0C	01	16	1B	38	35	22	2F
F	DC	D1	C6	CB	E8	E5	F2	FF

	8	9	A	B	C	D	E	F
0	68	65	72	7F	5C	51	46	4B
1	B8	B5	A2	AF	8C	81	96	9B
2	D3	DE	C9	C4	E7	EA	FD	F0
3	03	0E	19	14	37	3A	2D	20
4	05	08	1F	12	31	3C	2B	26
5	D5	D8	CF	C2	E1	EC	FB	F6
6	BE	B3	A4	A9	8A	87	90	9D
7	6E	63	74	79	5A	57	40	4D
8	B2	BF	A8	A5	86	8B	9C	91
9	62	6F	78	75	56	5B	4C	41
A	09	04	13	1E	3D	30	27	2A
B	D9	D4	C3	CE	ED	E0	F7	FA
C	DF	D2	C5	C8	EB	E6	F1	FC
D	0F	02	15	18	3B	36	21	2C
E	64	69	7E	73	50	5D	4A	47
F	B4	B9	AE	A3	80	8D	9A	97

Table D-4. LUT_09 Utilised During Decryption

	0	1	2	3	4	5	6	7
0	00	09	12	1B	24	2D	36	3F
1	48	41	5A	53	6C	65	7E	77
2	90	99	82	8B	B4	BD	A6	AF
3	D8	D1	CA	C3	FC	F5	EE	E7
4	3B	32	29	20	1F	16	0D	04
5	73	7A	61	68	57	5E	45	4C
6	AB	A2	B9	B0	8F	86	9D	94
7	E3	EA	F1	F8	C7	CE	D5	DC
8	76	7F	64	6D	52	5B	40	49
9	3E	37	2C	25	1A	13	08	01
A	E6	EF	F4	FD	C2	CB	D0	D9
B	AE	A7	BC	B5	8A	83	98	91
C	4D	44	5F	56	69	60	7B	72
D	05	0C	17	1E	21	28	33	3A
E	DD	D4	CF	C6	F9	F0	EB	E2
F	95	9C	87	8E	B1	B8	A3	AA

	8	9	A	B	C	D	E	F
0	EC	E5	FE	F7	C8	C1	DA	D3
1	A4	AD	B6	BF	80	89	92	9B
2	7C	75	6E	67	58	51	4A	43
3	34	3D	26	2F	10	19	02	0B
4	D7	DE	C5	CC	F3	FA	E1	E8
5	9F	96	8D	84	BB	B2	A9	A0
6	47	4E	55	5C	63	6A	71	78
7	0F	06	1D	14	2B	22	39	30
8	9A	93	88	81	BE	B7	AC	A5
9	D2	DB	C0	C9	F6	FF	E4	ED
A	0A	03	18	11	2E	27	3C	35
B	42	4B	50	59	66	6F	74	7D
C	A1	A8	B3	BA	85	8C	97	9E
D	E9	E0	FB	F2	CD	C4	DF	D6
E	31	38	23	2A	15	1C	07	0E
F	79	70	6B	62	5D	54	4F	46

Table D-5. LUT_0B Utilised During Decryption

	0	1	2	3	4	5	6	7
0	00	0B	16	1D	2C	27	3A	31
1	58	53	4E	45	74	7F	62	69
2	B0	BB	A6	AD	9C	97	8A	81
3	E8	E3	FE	F5	C4	CF	D2	D9
4	7B	70	6D	66	57	5C	41	4A
5	23	28	35	3E	0F	04	19	12
6	CB	C0	DD	D6	E7	EC	F1	FA
7	93	98	85	8E	BF	B4	A9	42
8	F6	FD	E0	EB	DA	D1	CC	C7
9	AE	A5	B8	B3	82	89	94	9F
A	46	4D	50	5B	6A	61	7C	77
B	1E	15	08	03	32	39	24	2F
C	8D	86	9B	90	A1	AA	B7	BC
D	D5	DE	C3	C8	F9	F2	EF	E4
E	3D	36	2B	20	11	1A	07	0C
F	65	6E	73	78	49	42	5F	54

	8	9	A	B	C	D	E	F
0	F7	FC	E1	EA	DB	D0	CD	C6
1	AF	A4	B9	B2	83	88	95	9E
2	47	4C	51	5A	6B	60	7D	76
3	1F	14	09	02	33	38	25	2E
4	8C	87	9A	91	A0	AB	B6	BD
5	D4	DF	C2	C9	F8	F3	EE	E5
6	3C	37	2A	21	10	1B	06	0D
7	64	6F	72	79	48	43	5E	55
8	01	0A	17	1C	2D	26	3B	30
9	59	52	4F	44	75	7E	63	68
A	B1	BA	A7	AC	9D	96	8B	80
B	E9	E2	FF	F4	C5	CE	D3	D8
C	7A	71	6C	67	56	5D	40	4B
D	22	29	34	3F	0E	05	18	13
E	CA	C1	DC	D7	E6	ED	F0	FB
F	92	99	84	8F	BE	B5	A8	A3

Table D-6. LUT_0E Utilised During Decryption

	0	1	2	3	4	5	6	7
0	00	0E	1C	12	38	36	24	2A
1	70	7E	6C	62	48	46	54	5A
2	E0	EE	FC	F2	D8	D6	C4	CA
3	90	9E	8C	82	A6	A8	B4	BA
4	DB	D5	C7	C9	E3	ED	FF	F1
5	AB	A5	B7	B9	93	9D	8F	81
6	3B	35	27	29	03	0D	1F	11
7	4B	45	57	59	73	7D	6F	61
8	AD	A3	B1	BF	95	9B	89	87
9	DD	D3	C1	CF	E5	EB	F9	F7
A	4D	43	51	5F	75	7B	69	67
B	3D	33	21	2F	05	0B	19	17
C	76	78	6A	64	4E	40	52	5C
D	06	08	1A	14	3E	30	22	2C
E	96	98	8A	84	AE	A0	B2	BC
F	E6	E8	FA	F4	DE	D0	C2	CC

	8	9	A	B	C	D	E	F
0	41	4F	5D	53	79	77	65	6B
1	31	3F	2D	23	09	07	15	1B
2	A1	AF	BD	B3	99	97	85	8B
3	D1	DF	CD	C3	E9	E7	F5	FB
4	9A	94	86	88	A2	AC	BE	B0
5	EA	E4	F9	F8	D2	DC	CE	C0
6	7A	74	66	68	42	4C	5E	50
7	0A	04	16	18	32	3C	2E	20
8	EC	E2	F0	FE	D4	DA	C8	C6
9	9C	92	80	8E	A4	AA	B8	B6
A	0C	02	10	1E	34	3A	28	26
B	7C	72	60	6E	44	4A	58	56
C	37	39	2B	25	0F	01	13	1D
D	47	49	5B	55	7F	71	63	6D
E	D7	D9	CB	C5	EF	E1	F3	FD
F	A7	A9	BB	B5	9F	91	83	8D

Appendix E – SHA-384/SHA-512 Constants

The 64-bit constants specified in the SHA-512 and SHA-384 specifications [107] are outlined below.

K_0 = 428a2f98 d728ae22		K_{20} = 2de92c6f 592b0275
K_1 = 71374491 23ef65cd		K_{21} = 4a7484aa 6ea6e483
K_2 = b5c0fbcf ec4d3b2f		K_{22} = 5cb0a9dc bd41fbd4
K_3 = e9b5dba5 8189dbbc		K_{23} = 76f988da 831153b5
K_4 = 3956c25b f348b538		K_{24} = 983e5152 ee66dfab
K_5 = 59f111f1 b605d019		K_{25} = a831c66d 2db43210
K_6 = 923f82a4 af194f9b		K_{26} = b00327c8 98fb213f
K_7 = ab1c5ed5 da6d8118		K_{27} = bf597fc7 beef0ee4
K_8 = d807aa98 a3030242		K_{28} = c6e00bf3 3da88fc2
K_9 = 12835b01 45706fbe		K_{29} = d5a79147 930aa725
K_{10} = 243185be 4ee4b28c		K_{30} = 06ca6351 e003826f
K_{11} = 550c7dc3 d5ffb4e2		K_{31} = 14292967 0a0e6e70
K_{12} = 72be5d74 f27b896f		K_{32} = 27b70a85 46d22ffc
K_{13} = 80deb1fe 3b1696b1		K_{33} = 2e1b2138 5c26c926
K_{14} = 9bdc06a7 25c71235		K_{34} = 4d2c6dfc 5ac42aed
K_{15} = c19bf174 cf692694		K_{35} = 53380d13 9d95b3df
K_{16} = e49b69c1 9ef14ad2		K_{36} = 650a7354 8baf63de
K_{17} = efbe4786 384f25e3		K_{37} = 766a0abb 3c77b2a8
K_{18} = 0fc19dc6 8b8cd5b5		K_{38} = 81c2c92e 47edaee6
K_{19} = 240ca1cc 77ac9c65		K_{39} = 92722c85 1482353b

K_{40} = a2bfe8a1 4cf10364
K_{41} = a81a664b bc423001
K_{42} = c24b8b70 d0f89791
K_{43} = c76c51a3 0654be30
K_{44} = d192e819 d6ef5218
K_{45} = d6990624 5565a910
K_{46} = f40e3585 5771202a
K_{47} = 106aa070 32bbd1b8
K_{48} = 19a4c116 b8d2d0c8
K_{49} = 1e376c08 5141ab53
K_{50} = 2748774c df8eeb99
K_{51} = 34b0bcb5 e19b48a8
K_{52} = 391c0cb3 c5c95a63
K_{53} = 4ed8aa4a e3418acb
K_{54} = 5b9cca4f 7763e373
K_{55} = 682e6ff3 d6b2b8a3
K_{56} = 748f82ee 5defb2fc
K_{57} = 78a5636f 43172f60
K_{58} = 84c87814 a1f0ab72
K_{59} = 8cc70208 1a6439ec

K_{60} = 90befffa 23631e28
K_{61} = a4506ceb de82bde9
K_{62} = bef9a3f7 b2c67915
K_{63} = c67178f2 e372532b
K_{64} = ca273ece ea26619c
K_{65} = d186b8c7 21c0c207
K_{66} = eada7dd6 cde0eb1e
K_{67} = f57d4f7f ee6ed178
K_{68} = 06f067aa 72176fba
K_{69} = 0a637dc5 a2c898a6
K_{70} = 113f9804 bef90dae
K_{71} = 1b710b35 131c471b
K_{72} = 28db77f5 23047d84
K_{73} = 32caab7b 40c72493
K_{74} = 3c9ebe0a 15c9bebc
K_{75} = 431d67c4 9c100d4c
K_{76} = 4cc5d4be cb3e42b6
K_{77} = 597f299c fc657e2a
K_{78} = 5fcb6fab 3ad6faec
K_{79} = 6c44198c 4a475817

References

[1] Diffie, W., Hellman, M. (1976), New Directions in Cryptography, IEEE Transactions on Information Theory, pp 664 – 654.
[2] Caloyannides, M. (2000), Encryption Wars: Early Battles, IEEE Spectrum, Volume 37, Number 4, April.
[3] SSH (2002), Introduction to Cryptography, URL: http://www.ssh.com/tech/crypto/.
[4] Singh, S. (1999), *The Code Book*, Fourth Estate Limited.
[5] Schneier, B. (1996), *Applied Cryptography – Protocols, Algorithms and Source Code in C*, John Wiley & Sons, Inc., 2nd Edition.
[6] Mollin, R.A. (2001), *An Introduction to Cryptography*, Chapman & Hall/CRC.
[7] Eskicioglu, A. and Litwin, L. (2001), Cryptography, IEEE Potentials, February/March.
[8] Pfleeger, C. (2000), *Security in Computing*, Second Edition, Prentice-Hall.
[9] Stallings, W. (1999), *Cryptography and Network Security Principles and Practice*, Second Edition, Prentice Hall International.
[10] Stinson, D. (1995), *Cryptography-Theory and Practice*, CRC Press.
[11] Delfs, H., Knebl, H. (2002), *Introduction to Cryptography – Principles and Practice*, Springer-Verlag.
[12] Brown, L. (1999), A Current Perspective on Encryption Algorithms, Uniforum NZ'99, New Zealand, April.
[13] Van Der Lubbe, J.C.A. (1998), *Basic Methods of Cryptography*, Cambridge University Press.
[14] Schneier, B. (2000), A Self-Study Course in Block Cipher Cryptanalysis, Cryptologia, Vol.24, No. 1, pp. 18-43, January.
[15] Kocher, P. (1996), Timing Attacks on Implementations of Diffie-Hellman, RSA, DSS and other Systems, Advances in Cryptology - CRYPTO'96, Springer-Verlag, LNCS 1109, ISBN 3-540-61512-1, pp 104-113.
[16] Kocher, P., Jaffe, J., Jun, B. (1999), Differential Power Analysis, Advances in Cryptology - CRYPTO'99, Springer-Verlag, LNCS 1666, ISBN 3-540-66347-9, pp 388-397.

[17] Elbirt, A.J., Yip, W., Chetwynd, B., Paar, C. (2000), An FPGA Implementation and Performance Evaluation of the AES Block Cipher Candidate Algorithm Finalists, The Third Advanced Encryption Standard (AES3) Candidate Conference, April, New York, USA.
[18] Ridge, D., Hamill, R., Craig, R., Farson, S., McCanny, J. (1996), Advanced DSP on FPGAs and CPLDs, ICSPAT - International Conference on Signal Processing Applications and Theory.
[19] Meyer-Baese, U. (2001), *Digital Signal Processing with Field Programmable Gate Arrays*, Springer-Verlag.
[20] Wade, W. (2001), Encryption Migrates to Silicon as Net Traffic Swells, EE Times, May.
[21] NCIPHER (2001), KPMG White Paper, URL:http://www.ncipher.com.
[22] Bohm, M. (2002), FPGA Evolution: New Design Methods on the Horizon, EE Times, January.
[23] Dandalis, A., Prasanna, V.K., Rolim, J.D.P. (2000), A Comparative Study of Performance of AES Candidates Using FPGAs, The Third Advanced Encryption Standard (AES3) Candidate Conference, April, New York, USA.
[24] Xilinx Virtex^{TM} FPGA Data Sheets (2001), URL: http://www.xilinx.com/partinfo/databook.htm.
[25] Curtin, M., Dolske, J. (1998), A Brute Force Search of DES Keyspace, The Magazine of USENIX and SAGE - ;login:, May.
[26] RSA Security (1997), Team of Universities, Companies and Individual Computer Users Linked Over the Internet Crack RSA's 56-Bit DES Challenge, URL: http://www.rsasecurity.com/news/pr/970619-1.html, June.
[27] Electronic Frontier Foundation (1999), Cracking DES: Secrets of Encryption Research, Wiretap Politics & Chip Design, URL: http://www.eff.org/descracker/.
[28] Nechvatal, J., Barker, E., Bassham, L., Burr, W., Dworkin, M., Foti, J., Roback, E. (2001), Report on the Development of the Advanced Encryption Standard (AES), Journal of Research of the National Institute of Standards and Technology, Volume 106, Number 2, URL: http://www.nist.gov/jres, May-June.
[29] Menezes, A., Oorschot, P., Vanstone, S. (1997), *Handbook of Applied Cryptography*, CRC Press.
[30] AES3 – Third AES Candidate Conference, (2000), URL: http://csrc.nist.gov/encryption/aes/round2/conf3/aes3conf.htm, New York.
[31] Gaj, K., Chodowiec, P. (2000), Comparison of the Hardware Performance of the AES Candidates using Reconfigurable Hardware, The Third Advanced Encryption Standard (AES3) Candidate Conference, April, New York, USA.
[32] Weeks, B., Bean, M., Rozylowicz, T., Ficke, C. (2000), Hardware Performance Simulations of Round 2 Advanced Encryption Standard Algorithms, The Third Advanced Encryption Standard (AES3) Candidate Conference, April, New York, USA.
[33] Ichikawa, T., Kasuya, T., Matsui, M. (2000), Hardware Evaluation of the AES Finalists, The Third Advanced Encryption Standard (AES3) Candidate Conference, April, New York, USA.
[34] Lidl, R., Niederreiter, H. (1994), *Introduction to Finite Fields and their Applications*, Cambridge University Press, Revised Edition, 1994.
[35] US National Institute of Standards and Technology - NIST (1999), Data Encryption Standard (DES), FIPS PUB 46-3, reaffirmed October.
[36] ANSI X9.17 (Revised) (1986), American National Standard for Financial Institution Key Management (Wholesale), American Bankers Association.
[37] ISO DIS 8732 (1987), Banking – Key Management (Wholesale), Association for Payment Clearing Services, London, December.

[38] McLoone, M., McCanny, J.V. (2000), A High Performance Implementation of DES, IEEE Workshop on Signal Processing Systems Design and Implementation - SiPS2000, Eds. M. Bayoumi, E. Freidman, IEEE Signal Processing Society Press, ISBN 0-7803-6488-0, pp374-383, Louisiana, USA, October.

[39] McLoone, M., McCanny, J.V. (2000), Data Encryption Apparatus, UK Patent Application, No. 0023409.6, Filed October.

[40] McLoone, M., McCanny, J.V. (2003), A High Performance FPGA Implementation of DES Using a Novel Method for Implementing the Key Schedule, IEE Proceedings – Circuits, Devices and Systems, accepted March 2003.

[41] National Bureau of Standards, (1980), DES Modes of Operation, Federal Information Processing Standards Publication, FIPS PUB 81, December.

[42] ANSI X9.52 (1998), Triple Data Encryption Algorithm Modes of Operation.

[43] Goubert, J., Hoornaert, F., Desmedt, Y. (1985), Efficient Hardware implementation of the DES, Advances in Cryptology – CRYPTO'84, Springer-Verlag, LNCS 196, ISBN 3-540-15658-5, pp 147-173, Berlin.

[44] Eberle, H. (1993), High-speed DES Implementation for Network Applications, 12th Annual International Cryptology Conference Proceedings - CRYPTO'92, Springer-Verlag, LNCS 0740, ISBN 3-540-57340-2, pp 521-539, California.

[45] Wilcox, D.C., Pierson, L.G., Robertson, P.J., Witzke, E.L., Gass, K. (1999), Sandia National Laboratories, A DES ASIC Suitable for Network Encryption at 10 Gps and Beyond, First International Workshop on Cryptographic Hardware and Embedded Systems - CHES '99, Springer-Verlag, LNCS 1717, ISBN 3-540-66646-X, pp 37-48, August.

[46] Leonard, J., Mangione-Smith, W.H. (1997), A Case Study of Partially Evaluated Hardware Circuits: Key-specific DES, Field Programmable Logic and Applications – FPL 1997, Springer-Verlag, LNCS 1304, ISBN 3-540-63465-7, pp151-160.

[47] Wong, K., Wark, M., Dawson, E. (1998), A Single-Chip FPGA Implementation of the Data Encryption Standard (DES) Algorithm, IEEE Globecom Communications Conference, pp 827-832, Piscataway, USA.

[48] Kaps, J.P., Paar, C. (1998), Fast DES Implementations for FPGAs and its Application to a Universal Key-Search Machine, 5th Annual Workshop on Selected Areas in Cryptography - SAC'98, Springer-Verlag, LNCS 1556, ISBN 3-540-65894-7, pp 234-247, Ontario, Canada, August.

[49] Patterson, C. (2000), Xilinx Inc., High Performance DES Encryption in Virtex FPGAs using Jbits, IEEE Symposium on Field-Programmable Custom Computing Machines - FCCM'00, IEEE Computer Society, ISBN 0-7695-0871-5, pp 113-121, California, April.

[50] Free-DES Core (2000), URL:http://www.free-ip.com/DES/, March.

[51] Trimberger, S., Pang, R., Singh, A. (2000), A 12Gpbs DES Encryptor/ Decryptor Core in an FPGA, Second International Workshop on Cryptographic Hardware and Embedded Systems - CHES 2000, Springer-Verlag, LNCS 1965, ISBN 3-540-41455-X, pp 156-163, August.

[52] Biham E. (1997), A Fast New DES Implementation in Software, Fast Software Encryption 4th International Workshop, FSE'97, Springer-Verlag, LNCS 1267, ISBN 3-540-63247-6, pp 260-271.

[53] Memec Design Services (1999), Alliance Core, XF-DES Data Encryption Standard Engine Core, URL:http://www.memecdesign.com/product, September.

[54] CAST, Inc. (1999), Alliance Core, X_DES Cryptoprocessor, URL:http://www.cast-inc.com/cores/xdes, February.

[55] Haskins, G.M. (1997), Securing Asynchronous Transfer Mode Networks, Masters thesis, Worcester Polytechnic Institute, Worcester, Massachusetts, USA, May.

[56] Daemen, J., Rijmen, V. (2002), *The Design of Rijndael: AES - The Advanced Encryption Standard*, Springer-Verlag, ISBN 3-540-42580-2.

[57] US National Institute of Standards and Technology (NIST) (2001), Advanced Encryption Standard, FIPS PUB 197, November.
[58] McLoone, M., McCanny, J.V. (2001), A Component for Generating Data Encryption/Decryption Apparatus, UK Patent Application No. 0111521.1, Filed May.
[59] McLoone, M., McCanny, J.V. (2001), High Performance Single-Chip FPGA Rijndael Algorithm Implementations, Third International Workshop on Cryptographic Hardware and Embedded Systems - CHES 2001, Eds. C Koç, D. Naccache, C. Paar, Springer-Verlag, ISBN 3-540-42521-7, pp65-77, Paris, France, May.
[60] McLoone, M., McCanny, J.V. (2001), High Performance FPGA Implementation of Rijndael, Irish Signals and Systems Conference - ISSC 2001, Eds. R. Shorten, T. Ward, T. Lysaght, ISBN 0-9015-1963-4, pp415-420, NUI Maynooth, Ireland, June.
[61] McLoone, M., McCanny, J.V. (2001), Single-Chip FPGA Implementation of Advanced Encryption Standard Algorithm, Field Programmable Logic and Applications - FPL 2001, Eds. G. Brebner, R. Woods, Springer-Verlag, ISBN 3-540-42499-7, pp152-161, Belfast, Northern Ireland, August.
[62] McLoone, M., McCanny, J.V. (2001), Apparatus for Selectably Encrypting and Decrypting Data", UK Patent Application No.0107592.8, Filed March.
[63] Chodowiec, P. Khuon, P., Gaj, K. (2001), Fast Implementations of Secret-Key Block Ciphers Using Mixed Inner- and Outer-Round Pipelining, Proc. 9th ACM International Symposium on Field-Programmable Gate Arrays- FPGA 2001, pp 94-102, California.
[64] McMillan, S., Patterson, C. (2001), Jbits^{TM} Implementations of the Advanced Encryption Standard (Rijndael), Field Programmable Logic and Applications - FPL 2001, Springer-Verlag, LNCS 2147, ISBN 3-540-42499-7, pp 162-171, Belfast, Northern Ireland, August.
[65] Mroczkowski, P. (2000), Implementation of the Block Cipher Rijndael Using Altera FPGA, The Third Advanced Encryption Standard (AES3) Candidate Conference, April, New York, USA.
[66] Gladman, B., (2001), The AES Algorithm (Rijndael) in C and C++, URL: http://fp.gladman.plus.com/cryptography_technology/rijndael/ index.htm, April.
[67] McLoone, M., McCanny, J.V. (2001), Rijndael FPGA Implementation Utilizing Look-Up Tables, IEEE Workshop on Signal Processing Systems Design and Implementation – SiPS 2001, Eds. F. Catthoor, M. Moonen, IEEE Signal Processing Society, ISBN 0-7803-7145-3, pp349-359, Antwerp, Belgium, September.
[68] McLoone, M., McCanny, J.V. (2003), Rijndael FPGA Implementations Utilizing Look-Up Tables, Journal of VLSI Signal Processing Systems, Eds. F. Catthoor, M. Moonen, Kluwer Academic Publishers, vol. 34-3, pp 261-275.
[69] McLoone, M., McCanny, J.V. (2003), Generic Architecture and Semiconductor IP cores for AES Cryptography, IEE Proceedings – Computers & Digital Techniques, accepted February 2003.
[70] Hu, Y., McLoone, M. (2001), An Apparatus Generating Encryption/Decryption Keys, UK Patent Application No.0121794.3, Filed September.
[71] Xilinx Virtex-II Pro^{TM} Platform FPGAs (2002), Introduction and Overview – Advanced Product Specifcation, URL: http://www.xilinx.com/publications/products/v2pro/ds_pdf/ds083-1.pdf
[72] US National Institute of Standards and Technology (NIST) (2001), Recommendation for Block Cipher Modes of Operation Methods and Techniques, Special Publication 800-38A, December.
[73] Bailey, D., Cammack, W., Guajardo, J., Paar, C. (1999), Cryptography in Modern Communication Systems, Texas Instruments DSPS FEST'99, Houston, August.
[74] Palnitkar, S. (1999), Implementing a Design Reuse Methodology Using a Design Data Management System, Synopsis User Group, SNUG'99, San Jose.
[75] Meiyappan, S., Jaramillo, K., Chambers, P. (1999), VHDL Coding Styles for Reusable, Synthesizable Designs, Synopsis User Group, SNUG'99, Boston.

[76] Stallings, W. (1995), *Network and Internetwork Security Principles and Practice*, Prentice Hall International.
[77] McLoone, M., McCanny, J.V. (2002), A Single-Chip IPSec Cryptographic Processor, IEEE Workshop on Signal Processing Systems Design and Implementation – SiPS 2002, ISBN 0-7803-7587-4, pp133-138, California, USA, October.
[78] McLoone, M., McCanny, J.V. (2002), Efficient Single-Chip Implementation of SHA-384 & SHA-512, IEEE International Conference on Field-Programmable Technology (FPT), ISBN 0-7803-7574-2, pp 311-314, Hong Kong, December.
[79] Cisco Systems Documentation (1999), Internetworking Technology Overview: Chapter 30 - Internet Protocols, URL: http//www.cisco.com, June.
[80] Frankel, S (2001), *Demystifying the IPSec Protocol*, Artech House Inc.
[81] Kent, S., Atkinson, R. (1998), Security Architecture for the Internet Protocol, RFC 2401, Internet Engineering Task Force (IETF), November.
[82] Kent, S., Atkinson, R. (1998), IP Authentication Header, RFC 2402, Internet Engineering Task Force (IETF), November.
[83] Kent, S., Atkinson, R. (1998), IP Encapsulating Security Payload (ESP), RFC 2406, Internet Engineering Task Force (IETF), November.
[84] Oppliger, R. (1997), *Internet and Intranet Security*, Artech House, Inc.
[85] Gollmann, D. (1999), *Computer Security*, John Wiley & Sons Ltd.
[86] Harkins, D., Carrel, D. (1998), The Internet Key Exchange (IKE), RFC 2409, Internet Engineering Task Force (IETF), November.
[87] Krawczyk, H., Bellare, M., Canetti, R. (1997), HMAC: Keyed-Hashing for Message Authentication, RFC2104, Internet Engineering Task Force (IETF), February.
[88] Madson, C., Glenn R. (1998), The Use of HMAC-SHA-1-96 within ESP and AH, RFC 2404, Internet Engineering Task Force (IETF), November.
[89] Ferguson, N., Schneier, B. (2000), A Cryptographic Evaluation of IPSec, Counterpane Internet Security Inc., http://www.counterpane.com/ipsec.html.
[90] Dandalis, A., Prasanna, V.K., Rolim, J.D.P. (2000b), An Adaptive Cryptographic Engine for IPSec Architectures, IEEE Symposium on Field-Programmable Custom Computing Machines - FCCM'00, IEEE Computer Society, ISBN 0-7695-0871-5, pp 132-144, California, April.
[91] Dobbertin, H. (1996), The Status of MD5 After a Recent Attack, RSA Laboratories' Crytobytes, vol 2, no. 2.
[92] US National Institute of Standards and Technology (NIST) (2001), Descriptions of SHA-256, SHA-384 and SHA-512, http://csrc.nist.gov/encryption/shs/sha256-384-512.pdf.
[93] US National Institute of Standards and Technology (NIST) (1995), Secure Hash Standard, FIPS PUB 180-1, April.
[94] US National Institute of Standards and Technology (NIST) (2002), The Keyed-Hash Message Authentication Code (HMAC), FIPS PUB 198, March.
[95] Pereira, R. (1998), IP Payload Compression Using DEFLATE, RFC 2394, Internet Engineering Task Force (IETF), November.
[96] McGregor, J.P., Lee, R.B. (2000), Performance Impact of Data Compression on Virtual Private Network Transactions, 25th IEEE Conference on Local Computer Networks – LCN 2000, IEEE Computer Society, ISBN 0-7695-0912-6, pp 500-510, November.
[97] Helion Technologies Limited (2002), Datasheet – High Performance SHA-1 Hash Core for Xilinx FPGA, URL: http://www.heliontech.com.
[98] Alma Technologies (2001), SHA-1 Core – Product Specification, URL: http://www.alma-tech.com.
[99] SETCo – SET Secure Electronic Transaction LLC (2002), SET Specification, http://www.setco.org/set_specifications.html

[100] Dierks, T., Allen, C. (1999), The TLS Protocol Version 1.0, RFC 2246, Internet Engineering Task Force (IETF), January.
[101] Chown, P. (2002), AES Ciphersuites for TLS, Internet draft, Internet Engineering Task Force (IETF), January.
[102] WAP – Wireless Application Protocol Forum (2001), Wireless Transport Layer Security (WTLS), WAP-261-WTLS-20010406-a, Version 06-Apr-2001, http://www.wapforum.org.
[103] RSA Security (2001), Improving Wireless LAN Authentication, www.rsasecurity.com.
[104] Borisov, N., Goldberg, I., Wagner, D. (2001), Intercepting Mobile Communications: The Insecurity of 802.11, 7th International Conference on Mobile Computing and Networking, Rome, Italy, July.
[105] Stubblefield, A. Ioannidis, J. Rubin, A. (2002), Using the Fluhrer, Mantin and Shamir Attack to Break WEP, Network and Distributed System Security Symposium - NDSS 2002, The Internet Society, California, February.
[106] Ericsson (2001), Wireless LAN Solution White-paper, www.ericsson.com.
[107] US National Institute of Standards and Technology (NIST) (2001), Secure Hash Standard, Draft FIPS PUB 180-2, May.
[108] Secucore (2001), SHA-256 Core, URL: http://www.secucore.com.
[109] HDL Design House (2002), HCR SHA 256, URL: http://www.hdl-dh.com.
[110] International Engineering Consortium (IEC) (2002), Virtual Private Networks (VPNs), URL: http://www.iec.org/online/tutorials/vpn/.
[111] RSA Security (2002), How Fast is the RSA Algorithm?, URL:http://www.rsasecurity.com/rsalabs/faq/3-1-2.html.
[112] Bitan, S. (1998), Hardware Implementation of IPSec: Performance implications, Firstvpn.com White Paper, URL:http://www.firsvpn.com/papers/radguard/Ipsec.pdf.
[113] Menezes, A.J. (1993), *Elliptic Curve Public Key Cryptosystems*, Kluwer Academic Publishers.
[114] Jurisic, A., Menezes, A.J. (1997), Elliptic Curves and Cryptography, Certicom White Paper, April.
[115] Orlando, G., Paar, C., (2000), A High-Performance Reconfigurable Elliptic Curve Processor for $GF(2^m)$, Second International Workshop on Cryptographic Hardware and Embedded Systems - CHES 2000, Springer-Verlag, LNCS 1965, ISBN 3-540-41455-X, pp 41-56, August.

Index

Active attack, 9
AES Development, 13, 14
Asymmetric Cryptosystem, 5
Authentication, 99

Brute-force attack, 8

Caesar cipher, 3
Chosen-Ciphertext Attack, 9
Chosen-Plaintext Attack, 9
Ciphertext-Only Attack, 9
Classical encryption, 2
Confidentiality, 99
Counter Mode, 86
Cryptanalysis, 8

DES CBC Mode, 33
DES CFB Mode, 34
DES Decryption, 32
DES ECB Mode, 33
DES Key Scheduling, 31
DES OFB Mode, 34
DES S-boxes, 31
DESCHALL, 13
Digital Signatures, 7

Elliptic Curve Cryptography, 130

Feistel cipher, 15
Finite Field Mathematics, 21
Function f, 29

Hash Functions, 7
HMAC, 107, 110

Integrity, 99
Internet Key Exchange, 129
IP Authentication Header, 101
IP Encapsulating Security
 Payload, 103
IPSec, 100
Irreducible polynomial, 22
ISAKMP, 129
Iterated cipher, 15
Iterative Looping, 10

Kerkhoff's principle, 8
Known-Plaintext Attack, 9

Loop unrolling, 10

Man-in-the-Middle Attack, 9

MARS, 2, 7, 14, 15, 16, 18, 19, 20, 21, 27
Mechanical Encryption Devices, 4
Message Authentication Code, 7
Multiplicative inverse, 25, 27

Non-repudiation, 99

Oakley Key Exchange, 129
One-time pad, 6

Passive attack, 9
Permutation cipher, 3
Pipelining, 10
Polyalphabetic substitution cipher, 3
Polynomial representation, 21
Power Analysis, 10
Power Analysis Attack, 10
Product cipher, 15
Public Key algorithm, 5

RC6, 2, 7, 14, 15, 16, 18, 19, 20, 21, 27
Rcon Function, 68
Rem Function, 68
Replay Protection, 101
Reusability, 88
Rijndael CBC Mode, 85
Rijndael CFB Mode, 85
Rijndael ECB Mode, 85
Rijndael Key Schedule, 61
Rijndael OFB Mode, 85
Rijndael Round, 59
Rotor machines, 4

RotWord Function, 68
RSA, 5

Secure Electronic Transaction Protocol, 115
Secure Socket Layer Protocol, 115
Serpent, 2, 7, 14, 15, 17, 18, 19, 20, 21, 27
SHA-1, 105, 108, 114
SHA-384, 117, 119, 151
SHA-512, 117, 119, 151
Skeme Key Exchange, 129
Sub-Pipelining, 10
Substitution Cipher, 2
Substitution-Permutation cipher, 15
SubWord Function, 68
Symmetric algorithm, 6

TDEA. *See* Triple-DES
Timing Attack, 9
Transport Layer Security Protocol, 115
Transposition Cipher, 3
Triple-DES, 37
Twofish, 2, 7, 14, 15, 16, 18, 19, 20, 21, 27

Vernam cipher, 6
Vigenère cipher, 3

Wired Equivalent Privacy Protocol, 116
Wireless Transport Layer Security Protocol, 116